Of Moths and Men:
An Evolutionary
Tale

JUDITH HOOPER

Of Moths and Men:
An Evolutionary
Tale

THE UNTOLD STORY OF SCIENCE
AND THE PEPPERED MOTH

W • W • NORTON & COMPANY
NEW YORK • LONDON

Copyright © 2002 Judith Hooper
First American edition 2002

Library of Congress Cataloging-in-Publication Data
Judith Hooper.
Of moths and men : an evolutionary tale : the untold story of science and the peppered moth /
Judith Hooper.—1st American ed.
p. cm.
Includes bibliographical references (p.) and index.
ISBN 0-393-05121-8
1. Natural selection. 2. Evolution (Biology). 3. Peppered moth. 4. Fraud in science. I. Title.

QH375 .H66 2002
576.8'2'092—dc21
[B]   2002026315

W.W. Norton & Company, Inc.
500 Fifth Avenue, New York, NY 10110
www.norton.com

W.W. Norton & Company Ltd.
10 Coptic Street, London WCIA 1PU

1 2 3 4 5 6 7 8 9 0

For Dick and Jake Teresi

Contents

Acknowledgements

Without the help of many people who gave generously of their time it would have been impossible to conjure this story out of the past. In the UK I am especially indebted to Laurence and Merrill Cook, and to John and Sandra Turner, who were hospitable and helpful beyond the call of duty. Kate Davies was a godsend. Much of what I learned of the world of E.B. Ford and H.B.D. Kettlewell came from the vivid stories and keen perceptions of Kate Davies, R.J. (Sam) Berry, Dame Miriam Rothschild, Laurence Cook, John Turner, Bryan Clarke, David Lees, Kauri Mikkola, James Cadbury, John Haywood, Ruth Wickett and Audrey Smith. Michael Majerus spent hours teaching me engagingly about peppered moths.

On this side of the Atlantic I am grateful, first of all, to Ted Sargent, for being willing to share his personal story so candidly and for fielding countless questions with great clarity. Lincoln Brower, David Jones and Jim Murray were mother lodes of information about the Oxford School of Ecological Genetics in the 1950s and in many cases bent over backwards to look up facts and locate ancillary source material. In matters scientific I am indebted to Bruce Grant, John Endler, Jerry Coyne, Richard Harrison and Douglas Futuyma. Will Provine allowed me ramble through his library and his expansive mind. Betty Smocovitis and Egbert Leigh pointed out salient historical themes. Don Kroodsma lucidly analysed the experimental design of Kettlewell's classic experiments, and Austin (Bob) Platt, in addition to sharing his memories, conscientiously read over the manuscript for accuracy. I am very grateful to Ernst Mayr and Lynn Margulis for reading over the book and providing valuable insights.

David Kettlewell, who, sadly, died suddenly in October 2000, added an invaluable dimension to the book. Thanks to Chris Carlisle, who tipped me off to the peppered moth controversy in the first place, to Craig Holdrege, whose article started this journey and to Bob Crease for memorable phrases. Miranda Kettlewell kindly provided photographs. Jim Kettlewell added a delightful story. Marian Dowdeswell and Dr Erasmus Barlow offered much needed assistance. Thanks to Lynsey Beauchamp and my friends at the British Meditation Society for their TLC.

My research owes much to Adrian Hale at the Wolfson College Library; to the helpful and efficient staff of the Western Manuscripts Reading Room of the Bodleian Library; to Rob Cox of the American Philosophical Society library; and to the libraries of Amherst College and the University of Massachusetts. Steve Jebson of the National Meteorological Library and Archives kindly provided weather records.

Finally, I was fortunate to have my words fall into the hands of two unusually deft, perspicacious and dedicated editors, Bob Weil at W.W. Norton and Clive Priddle at 4th Estate, and to enjoy the unwavering support and delightful humour of my agent, Michael Carlisle, who believed in moths from the start. Thanks also to Steve Cox for his admirable copy-editing.

He told me about the odors of butterflies — musk and vanilla; about the voices of butterflies; about the piercing sound given out by the monstrous caterpillar of a Malayan hawkmoth, an improvement on the mouselike squeak of our Death's Head moth; about the small resonant tympanum of certain tiger moths; about the cunning butterfly in the Brazilian forest which imitates the whir of a local bird. He told me about the incredible artistic wit of mimetic disguise . . .

Vladimir Nabokov, *The Gift*

In questions of science, the authority of a thousand is not worth the humble reasoning of a single individual.

Galileo Galilei

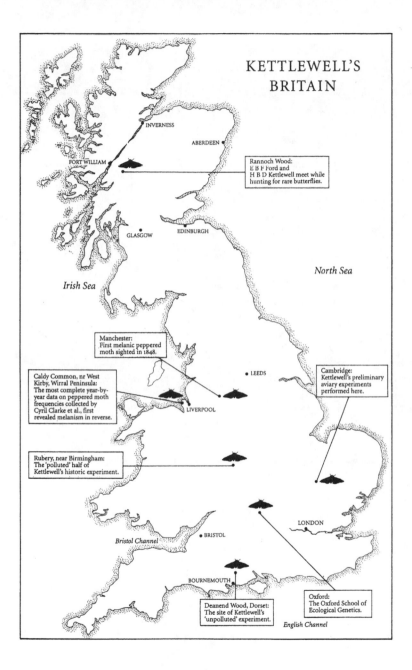

KETTLEWELL'S BRITAIN

INVERNESS

ABERDEEN

Rannoch Wood:
E B F Ford and
H B D Kettlewell meet while
hunting for rare butterflies.

FORT WILLIAM

GLASGOW   EDINBURGH

North Sea

Irish Sea

Manchester:
First melanic peppered
moth sighted in 1848.

Caldy Common, nr West
Kirby, Wirral Peninsula:
The most complete year-by-
year data on peppered moth
frequencies collected by
Cyril Clarke et al., first
revealed melanism in reverse.

• LEEDS

Cambridge:
Kettlewell's preliminary
aviary experiments
performed here.

LIVERPOOL

Rubery, near Birmingham:
The 'polluted' half of
Kettlewell's historic experiment.

LONDON

Bristol Channel   • BRISTOL

BOURNEMOUTH

Deanend Wood, Dorset:
The site of Kettlewell's
'unpolluted' experiment.

Oxford:
The Oxford School of
Ecological Genetics.

English Channel

Prologue:
The moths of Oxford

The moths of this story – the peppered moths of England – became the most famous insects in the world. You might say they were at the right place at the right time. If you have studied biology in school or college you will have glimpsed them in textbook photographs, posed on tree trunks, immortalized and still as figures in a classical frieze. They were, and still are, hailed as 'Darwin's missing evidence', evolution's 'prize horse', 'evolution in action'. Yet the history of the peppered moth has lately become a battlefield, the controversy growing more inflamed by the moment. At first glance it is hard to imagine how so much passion, intrigue and even tragedy could attach to the fate of a population of insects.

Under the surface of the bare-bones tale allotted to the peppered moth in every textbook, there lies, like a fading palimpsest, a human story, set in the little-known community of moth collectors. In fact, this unlikely protagonist – a mere moth – has engendered a century of struggle, for, contrary to popular wisdom, Darwin's theory of natural selection was on shaky ground in the first part of the twentieth century. Apart from admittedly artificial laboratory studies of fruit flies, experimental evidence for Darwinian theory was lacking. There was doubt that natural selection could ever be experimentally demonstrated at all, and if it could not then evolutionary biology was doomed to be a second-rate 'historical' science, incapable of ever being proved or disproved. Fighting against a sea of Lamarckians, saltationists, macro-mutationists, and 'random drift' believers, a resolute band of neo-Darwinists sought the perfect example to prove their case.

The man who would deliver this impeccable proof was an unlikely scientific hero. Dr H.B.D. Kettlewell was not a research scientist by training but a medical doctor, a charismatic and volatile amateur lepidopterist. He had been recruited to Oxford by E.B. Ford, famous geneticist, eccentric don, and megalomaniac founder of the Oxford School of Ecological Genetics. By his own lights, Ford had almost single-handedly rescued natural selection from oblivion in the 1920s and 1930s, when convinced Darwinists were few, and throughout his life he headed a scientific coterie with very definite ideas about the way evolution worked. All of these ideas were absorbed by Kettlewell and taken into the field. For this and other reasons, the moths in his experiments, the moths of the photographs, had to support a heavy freight of intellectual baggage.

The night-flying peppered moth, *Biston betularia*, normally off-white with a freckling of black scales, had been an object of scientific curiosity since Darwin's own time. In the mid-nineteenth century a dark ('melanic') form of this moth appeared in the industrialized Midlands of the British Isles. These strange black moths thrived and grew more numerous wherever there was factory smoke fouling the atmosphere, and some people began to think that Darwin's theory of evolution might explain why. Resting against soot-blackened backgrounds the melanic moths were nearly invisible to birds, and so escaped being preyed upon. Thus more of them survived to reproduce. In rural areas, it was just the opposite. There it was the normal, speckled variety of the peppered moth that had the advantage, well camouflaged against the light, lichen-covered tree trunks. In Darwinian language, natural selection favoured the black appearance in the grimy mill towns and the light one in rural, unpolluted woodlands. The camouflaged moths in each case were the 'fittest', surviving and reproducing in greater numbers than their poorly camouflaged brethren.

For fifty years, this interpretation of industrial melanism, as it was known, was only a theory – until Kettlewell ventured out into the English countryside to prove it in 1953. Here is one textbook summary of his experiments:

In carefully controlled experiments, equal numbers of peppered and melanic moths were released into the countryside where industry was absent. The trunks of the region are covered by lightly colored lichens. After release, observers with binoculars tried to determine the fate of as many moths as possible ... Of the 190 moths that were actually observed to be eaten by the birds, 26 were peppered [pale] and 164 were melanic ...

The same experiment was conducted in the midland of England, where extensive industrialization had caused most lichens to die and trees became darkened by soot. The results of this experiment were the reverse of the previous experiment. The birds ate many more peppered moths than melanic moths ...[1]

Neat and tidy mirror-images, these twin demonstrations became the most celebrated experiment ever in evolutionary biology. By the 1960s it had conquered all the textbooks, influencing the minds of four decades of biology students. It is the slam dunk of natural selection, the paradigmatic story that converts high school and college students to Darwin, the thundering left hook to the jaw of creationism.

*   *   *

Craig Holdrege, a young American biology teacher, had been teaching the peppered moth story for several years. In 1986 he was reading a paper by Sir Cyril Clarke, a respected British peppered-moth researcher, when an offhand comment hit him like a spray of arctic water. Casually, Sir Cyril, who had been a bosom friend of Kettlewell, stated: 'All we have observed is where the moths do not spend the day. In 25 years we have only found two *betularia* on the tree trunks or walls adjacent to our traps . . .'

'What is going on here?' Holdrege asked himself. He had been displaying photographs of moths on tree trunks, telling his students about birds selectively picking off the conspicuous ones . . . 'And now someone who has researched the moth for 25 years reports having seen only two moths sitting on tree trunks.'[2] What about the lichens, the soot, the camouflage, the birds? What about the grand story of industrial melanism? Didn't it depend on moths habitually resting on tree trunks?

Unnerved by Clarke's remark, Holdrege dug up H.B.D. Kettlewell's original journal articles, which very few biology teachers had actually read. He was appalled. As his private quest took him to other papers and other scientists, he began to realize that the standard theory of industrial melanism, treated as gospel in the textbooks, was full of holes.

As it turned out, Holdrege was not the only one to notice the cracks in the icon. Before long the peppered moth had kindled a smouldering scientific feud, at first conducted in the footnotes of journals and, later, in textbook wars, away from the public eye.

The ferocity of the debate was puzzling and intriguing to me. The peppered moth story seemed fairly straightforward, and if parts of it were wrong, or even most of it, so what? Surely there were other instances to demonstrate evolution? What was it about this one case that was so sacred? Perhaps it was a symbol of something, a vital totem, and that was why some people were defending it so tenaciously. This book grew out of those questions.

Behind the story, like a monster lurking under a five-year-old's bed, is the bogeyman of creationism. Worried friends have asked me: 'Aren't you playing into the hands of creationists?' The truth is that it is already too late to disguise the damage. Since the first scent of publicized trouble in late 1999, the peppered moth has been discussed *ad nauseam* on Christian websites as if it were an issue of theology, like the trinitarian doctrine. 'If the *best* example of evolution that has been believed and used as proof by eminent evolutionary biologists for over 40 years is not true, how about all their *other* supposed examples? It makes you wonder, doesn't it?' proclaims a website associated with the Crystal Lake Church of Christ. The paragraphs allotted to the experiments in textbooks have become zones of intellectual trench warfare, causing the atmosphere in the biological sciences to become so polarized that almost any criticism is viewed as creationist fodder. However, as one researcher I interviewed put it. 'You can't stop doing science just because you're worried about the creationists.' For the record, I am not a creationist, but to be uncritical about science is to make it into a dogma.

The roots of the peppered moth story lie deep in the personal histories of a small number of scientists, preserved in the memories of friends, colleagues and family, as well as in letters bound in satin ribbons and archived at the New Bodleian Library at Oxford. At the very moment that the peppered moth experiments were establishing the Oxford biologists as masters of their world, their personal and professional relationships were disintegrating in a miasma of recriminations, intrigue, jealousy, back-stabbing and shattered dreams. They conceived the evidence that would carry the vital intellectual argument, but at its core lay flawed science, dubious methodology, and wishful thinking. Clustered around the peppered moth is a swarm of human ambitions, and self-delusions shared among some of the most renowned evolutionary biologists of our era.

PART I

I

The moonlit world

To begin at the beginning, the Lepidoptera are divided into two orders: butterflies (*Rhopalocera*) and moths (*Heterocera*). The rule of thumb for distinguishing between them is that butterflies have antennae with little knobs at the tip, while those of moths are threadlike, spindlelike or feathery – though there are exceptions. A second rule of thumb is: 'If it's not a butterfly, it's a moth.' Some might say that butterflies are Lepidoptera that people like while moths are ones they don't like, and yet as nocturnal flyers, linked to darkness and phases of the moon, moths are in many ways more romantic and mysterious than their diurnal cousins.

While butterflies are netted in sunny meadows, most moths must be caught after sundown using either light traps or pheromone traps baited with female moths. There are plenty of female butterfly collectors, but most moth devotees seem to be male. One celebrated example is W.J. Holland, author in 1903 of *The Moth Book*, which though outdated in many respects is still a bible for collectors. Its colour plates quickly disabused me of any prejudices I had about moth drabness. Here were rows and rows of moths set out like gems: mysterious underwings with flamboyant undergarments, splendid and eerie cecropia moths, startling luna moths with 'eyes' etched on their lime-green wings, elegant shocking-pink hawkmoths, tiger moths clothed in intricate wallpaper-like patterns. One of the stylistically most vivid passages in the book describes the mystique of the nocturnal world:

There are two worlds, the world of sunshine and the world of the dark. Most of us are more or less familiarly acquainted with the first; very few of us well acquainted with the latter. Our eyes are well adapted to serve us in the daylight, but they do not serve us as well in the dark ... There are whole armies of living things which, when we go to sleep, begin to awaken ...

Among the insects thousands and tens of thousands are nocturnal ... When the hour of dusk approaches, stand by a bed of evening primroses, and, as their great yellow blossoms suddenly open, watch the hawkmoths coming as swiftly as meteors through the air, hovering for an instant over this blossom, probing into the sweet depths of another, and then dashing off again so quickly that the eye cannot follow them ...

The Moth Book was instrumental in changing Ted Sargent's career. He was still a bird behaviourist when he took a post in 1963 at the University of Massachusetts at Amherst. A picture-postcard town with a garrulously populist 'town meeting' government and an old-fashioned New England common at its heart, Amherst remained rural enough to have dairy farms within the city limits, where the lowing of cows mingled with the meditations of Dante scholars. The surrounding 'hill towns' were more bucolic still. In quiet Pelham, a town with a single tiny grocery store, Sargent, then a bachelor assistant professor, rented rooms in a house with a large, leafy backyard, and one humid summer evening he tried an old trick described by Holland in *The Moth Book*: mix brown sugar with beer, paint the solution on tree trunks – this is called 'sugaring' – and when night falls, rush out with a jar and a light. Moths will be found all over the trees, ready to be scooped up.

'Holland describes the excitement,' says Sargent. 'And it *is* exciting. You run around picking the moths off the trees. Their eyes shine like coals. The brown sugar and beer smells like fermenting fruit. Some people use cat piss instead of beer. Everyone has his secret recipe.'

The British lepidopterist P.B.M. Allan, author of the popular *Moth Hunter's Gossip* (1937), always insisted that the only foolproof sugaring ingredient was brown Barbados sugar; no other kind would do. The most renowned literary lepidopterist of the twentieth century, Vladimir Nabokov, had his own recipe: molasses, beer, and rum. 'Through the gusty blackness,' he wrote in *Speak, Memory*, 'one's lantern would illumine the stickily glistening furrows of the bark and two or three large moths upon it imbibing the sweets, their nervous wings half open butterfly fashion, the lower ones exhibiting their incredibly crimson silk form beneath the lichen-gray primaries.' Nabokov's moth sounds as if it must have been an underwing, a moth that wears a dull grey or brown skirt over a startlingly colourful underskirt.

In time, using his own varieties of sugar solutions, Sargent identified some forty underwing species in his backyard and became so infatuated that he put aside his bird studies. He had been feeling stale anyway, shut up in the lab, studying caged birds. 'Moths are just so beautiful and there are so many of them,' he noted. Switching to moths, he became an expert on the *Catocala* (underwing) genus, but in the process he also got hooked on the mystery of melanic (black) moths.

Apart from an unsuccessful stint with a butterfly net at age seven that bagged only a few enraged hornets, I myself had paid scant attention to butterflies and even less to moths. I found it hard to fathom why so much passion would attach to an order of bothersome insects whose only purpose seemed to be to slip through holes in the screen to immolate themselves upon my reading lamp. Their stumpy, brownish bodies and kamikaze ways repelled me, and, had I taken the time to look closely, so would the feathery antennae and the space-alien honeycombed eyes. Against my skin the frantic, powdery wings felt like cobwebs, and I'd brush them off with an impatient hand, never giving a thought to their small lives: their lifespan, what they ate, how they mated, how they survived, why exactly they had this compulsion to career into a hot 60-watt bulb. I had, in fact, never really *seen* them at all, a fact that was impressed on me forcibly by one Jean Henri Cassirer Fabre (1823–1915), who in his 1915 book *The Life of a Caterpillar* described the emergence of a moth called *Psyche*:

In their modest pearl-grey dress, with their insignificant wing-equipment, hardly exceeding that of a Common Fly, our little Moths are still not without elegance. They have handsome feathery plumes for antennae. Their wings are edged with delicate fringes. They whirl very fussily inside the bell-jar; they skim the ground, fluttering their wings; they crowd eagerly around certain sheaths which nothing on the outside distinguishes from the others. They alight upon them and sound them with their plumes.

This feverish agitation marks them as lovers in search of their brides. This one here, that one there, each of them finds his mate. But the coy one does not leave her home. Things happen very discreetly through the wicket left open at the free end of the case. The male stands on the threshold of this back-door for a little while; and then it is over; the wedding is finished. There is no need for us to linger over these nuptials in which the parties concerned do not know, do not see each other.[3]

Ruminating lavishly on each event of the insect world as if it were a dinner *chez* Guermantes, the strange and solitary Fabre converted hordes of schoolboys to entomology. Such was the allure of natural history that his fame in his time – certainly in his native France – rivalled Darwin's. At his country home in Provence, where he also did seminal work in botany, the 'Homer of Insects' prefigured the modern discipline of animal behaviour by devising numerous experiments, such as attaching tiny burdens to beetles' backs to see how heavy a load they could support, or shooting a cannon to test the hearing of cicadas.

During his remarkably long life Fabre was an intimate of John Stuart Mill and attracted the admiration of Darwin, Poincaré, Mallarmé, Maeterlinck, and Bergson, yet in a modern setting he might have come across as the sort of marginal fanatic to be avoided at all costs. It seems possible that the charm of entomology might include the faintly sadistic prospect of lording it over very tiny creatures that are completely in one's power.

Like Fabre, entomologists in general tend to be obsessed, as they freely admit. When asked for his views of the Creator, the biologist J.B.S. Haldane quipped that he could deduce only that He must have an excessive fondness for beetles, an in-joke for entomologists, for there are more kinds of beetles on Earth than any other animal – about 500,000 species, a prodigality of nature that attracts squadrons of fanatical beetle collectors, many equipped with expensive wooden carrying cases for their specimens. The young Charles Darwin was a typical nineteenth-century beetle fancier who crammed his rooms at Cambridge with the nut-hard exoskeletons; later, as a mature scientist, he would spend eight years studying barnacles. It is difficult for the rest of us to imagine the avidity of such a quest. It seems probable that lepidopterists are the most obsessed, the most besotted, of the lot.

'We are complete nut-cases,' says Luxembourg-born lepidopterist Michael Majerus, of Cambridge University. 'You spend many many nights walking around woodlands in England with a torch and a moth trap. Anyone who encountered you would think you were extraordinarily strange, but you are as happy as you could possibly be.' Lepidopterists say that it is a matter of temperament; you are either made for it or you're not. Part of having the right temperament is the ability to sit patiently for long periods doing nothing. 'It would drive many people bonkers going out every day at dawn, and very often you're sitting there watching very little for long periods of time,' Majerus adds. 'I spent one summer working on the clouded yellow butterfly, which only flies in sunshine in Swedish Lapland. Every morning I had to climb to eight hundred metres and sit down, and many days the sun never came out, and at four o'clock I would just walk down the mountain. It was one of the most idyllic months I ever spent.'

There are many more moths than butterflies in this world. The latter number only about 20,000 out of some 165,000 Lepidoptera in total. Consequently the most gifted moth connoisseurs must have a savant-like knowledge of the insects' habitat, range, and larval food plants, which they usually do not hesitate to display. In Massachusetts, where there are sixty species of butterfly and thousands of moths, even Ted Sargent is awed by enthusiasts who 'can go up to a streetlight and start naming these things . . . It's an extraordinary talent.'

Sargent has reached the conclusion that moth people are a breed apart, 'weirder and crazier than butterfly people. They have this male bonding thing; they have a language all their own, a latinized vocabulary, since the moths and the food plants have Latin names.' Like other rapt hobbyists – horse and dog breeders, experts on Chippendale chairs, conspiracy theorists perhaps – lepidopterists can be crushing bores if you're not one of them. In his lepidopterist writing, even the consummate stylist Vladimir Nabokov could be dense, fussy and jargon-ridden, with only the occasional turn of phrase ('cloudlet', skies of 'impeccable blue') to distinguish his postings from the usual fare. *The female of* sublivens *is of a curiously arctic appearance, completely different from the richly pigmented, regionally sympatric, locoweed- and alfalfa-feeding* L. melissa . . . And so on.

Some moth men have the stunted social skills of the more monomaniacal computer hackers, going about with misbuttoned shirts and uncombed hair, spouting taxonomic Latin. Their dinner-table small-talk is likely to be confined to adoring descriptions of a certain moth, its foodplant, its range, who first named it. *I think the such and such is extending its range. So and so reported* . . . They will go to extraordinary lengths to catch a rare specimen. One obsessed collector acquaintance of Sargent's goes into the field wearing a cap with a light mounted on it. From his belt dangle his killing jars, and secured in a holster at his waist sits a cylinder of sugaring solution the size of an oxygen tank. He stays at a cheap motel on a busy commercial strip and collects at night behind the motel, wandering into people's landscaped yards and spraying the trees with his sugaring solution.

*   *   *

It might be said that Lepidoptera are all about sex and death
– as indeed, from a Darwinian perspective, everything is. Like
butterflies, moths pass through four life-cycle stages: egg, larva
(caterpillar), pupa and adult (known as the 'perfect insect' or
'imago'). Technically, metamorphosis is a form of moulting,
which all animals with exoskeletons must undergo to grow.
The larvae exist only to eat, feeding (and feeding and feeding)
on their specific hostplants. Children who collect caterpillars
quickly learn that to keep the caterpillar alive you must keep
its foodplant in the box with it. The identity of the larval
foodplant of rare species is part of the lore of lepidoptery,
discussed endlessly in the journals. As they grow, caterpillars
undergo several distinct moults, or 'instars', before becoming
transformed into pupae, or chrysalids, which may be naked or
covered with a cocoon more or less composed of silk. After
spending a period of time in the pupal state, they appear
as four-winged, six-legged insects. Nabokov once described
their metamorphosis thus: 'There comes for every caterpillar
a difficult moment when he becomes pervaded by an odd
discomfort. It is a tight feeling – here about the neck and
elsewhere, and then an unbearable itch. Of course, he has
moulted a few times before, but that is nothing in comparison
to the tickle and urge that he feels now. He must shed that
tight, dry skin, or die.' Some species pupate underground;
others in nests. In some species it is the eggs that 'overwinter',
or hibernate; in others it is the larvae or pupae that do so.

Moth anatomy is as complex as it is tiny, requiring a battery
of lilliputian magnifying and surgical tools. Just consider the
legs: as in all insects, these consist of *coxa, trochanter, femur,
tibia* and *tarsus*, the latter composed of five joints and armed
with hooks or claws, known technically as the ungues. The
genitalia of Lepidoptera are usually dissected with care and
debated fiercely, as the fine points of their structure commonly

reveal the identity of the species. And all this pales next to the complexity of the breeding experiments undertaken by serious amateurs. 'In Britain alone 780 species of Macrolepidoptera . . . demand 780 different techniques in breeding in each of their four stages of metamorphosis,' noted Kettlewell.

A typical breeding handbook, this one translated from the German, throws the reader into a realm more demanding in its way than neurosurgery or nuclear physics. The sense of secret lore, the painstaking (and numbingly boring) instructions, the solitary, never-ending toil, is reminiscent of early alchemical treatises. Under the heading 'Arctiidae (Footmen, Tigers and Ermines)', for example, one reads such instructions as:

A large number of species overwinter as larvae, often after the second moult, in plucked moss. They must not be kept too dry and in mild winter weather they often become quite active and occasionally nibble brambly leaves or cabbage stalks. The winter-hardy blue-green saxifrage (*Saxifraga caesia*) is also readily accepted. The overwintering larvae may generally be brought into the warmth in March. If mating and oviposition have occurred earlier than in nature, the larvae begin to diapause equally early. In this case, it is more practical to keep the stock in an airy cage, containing fairly damp moss, at *c.* o C in a refrigerator for two or three months . . .[4]

If you feel like meditating on the brevity of earthly life, consider a moth. A sudden frost can murder thousands of eggs. Larvae perish by the million, from starvation, predators, bad weather or disease. Unspeakable viruses exist that are 100 per cent fatal to caterpillars, reducing their tissues to jelly. If you're a caterpillar in the wild your odds of dying before you become a moth are at least 90 per cent. Moth wings, tissue-paper-thin, can be easily torn. A moth or butterfly can be blown off-course by the wind. There is no protection against predators beyond one's fated coloration (or, in some cases, a noxious taste). Life according to Darwin's inexorable mathematics is cruel and brief. Even Darwin shuddered at nature's rapacity, which, by some accounts, convinced him of the nonexistence of God.

The adult moth's entire career is sex. Many moths and butterflies do not even have mouths, for there is no eating required during their few days of winged life. Their sole imperative is to mate, and, if female, to lay. 'They are,' wrote Holland, 'simply animate, winged reservoirs of reproductive energy, and, when the sexual functions have been completed, they die.' As the dewy new female moth emerges from her pupa she emits a pheromonal scent that summons the males of the species from miles away. Copulation ensues within minutes or hours and is of unusually long duration. As a boy in Wisconsin Paul Ewald, a biologist at Amherst College, kept moths in his backyard, not an unusual childhood hobby for people in his profession. At dusk, when the females emerged, or *eclosed* (another lovely lepidopterist word, suggestive of French poetry), squadrons of male moths would streak through the violet sky, over the backyard fence, into the cage. 'I was amazed at how long they stayed together,' Ewald told me, a remark that I at first took to refer to some kind of monogamy, as with wolves, until he explained that he meant the length of time the moths remained *in copula*. Many mating couples spend the whole night together, before flying away in the morning. (Active mating requires about a half hour; the remainder of the time is just 'resting'.)

Perhaps because of the intense associations they stir up in humans, there are many poems about moths, self-immolation by candle and transfiguration being prominent themes. Here is a typical eighteenth-century example, by Henry Brooke:

From death their happier life derive,
and tho' apparently entomb'd, revive:
Chang'd, thro; amazing transmigration rise,
And wing the regions of unwonted skies;
So late depress'd, contemptible on earth,
Now elevate to heaven by second birth.

Even the names of moths are poetic. In Ted Sargent's genus alone, there are moths called Inconsolable (*insolabilis*), Dejected (*dejecta*), the Tearful (*lachrymosa*), Serene (*serena*), the Sweetheart (*amatrix*), Bride (*neogama*), and penitent (*piatrix*). There is a Widow, a Forsaken, a Betrothed, a Consort, an Old Maid, and many other names that seem to allude to a distant romantic tragedy.[5] There are Lepidoptera called Pale Brindled Beauty, Mottled Beauty, Scalloped Hazel, Queen of Spain Fritillary, Handmaid, Lesser Treble-bar, Balsam Carpet. The ultimate entomological triumph is to name a genus, species or variety, and the rules governing this Adamic enterprise are spelled out by the International Code of Zoological Nomenclature. Once the name is published in an entomological journal, it may be listed in the annual Zoological Record. The genus comes first, followed by the species, then the variety. In the case of the peppered moth *Biston betularia* f. *typica, Biston* is the genus, *betularia* is the species, and the form, or variety, is *typica* (or alternatively f. *carbonaria*, or *insularia*, as we shall see). It is not uncommon for the discoverer to affix his own name, so that we have butterflies called *Eupithecia actaeata* Walderdorff.

Lepidoptera seem made for genetics because they are fairly easy to breed and raise and generally produce one generation a year. They also wear their genetic identity on their wings, and these patterns have an important evolutionary function. Many Lepidoptera depend for their survival on concealment alone. According to Bernard Kettlewell, who became the champion of the peppered moth, 'Their wing patterning, unlike that in smaller insects such as [the fruit fly] *Drosophila*, may be their most important character. Their gene-complex is, in part, written into the architecture of their wings. Night-flying moths in particular must, in order to survive, pass the day motionless and invisible on particular backgrounds.'

In other words, how well one's wings may resemble a pattern of twigs, a dirt clog, a leaf, another moth, or a piece of bark could spell whether one is eaten or not, whether one perishes or survives long enough to reproduce. For the peppered moths of England, this question of crypsis, or camouflage, will become the cynosure of a great Darwinian drama.

2

What Darwin missed

The typical form of *Biston betularia*, a member of the geometrid family (*Geometridae*), is cream-coloured, with a speckling of dark lines and spots. In photographs it looks dim and ordinary next to all the fancy moths with crimson hindwings, orange tiger spots or spangles of green iridescence. When I saw the real thing at Oxford's Hope Entomology Department, dead, pinned and displayed like a jewelled pendant against a dark velvety background, I was taken aback by its beauty. The dark markings etched on the pearly wings had the understated elegance of certain Japanese textiles, compared with which all those vivid tiger moths now seemed gaudy and immodest. I am speaking of the ordinary form, *Biston betularia*, f. *typica*. It is hard to see how the inky-black melanic form, *carbonaria*, could appear lovely to anyone except a nineteenth-century entomologist desperate for an exotic variant.

In 1848 a lepidopterist named R.L. Edleston caught and pinned a rare dark form of *Biston betularia*. He was not a pampered gentleman of leisure like many entomologists but a modest calico-maker who lived in a squat brick house in a dingy warren of dark streets and alleyways. Edleston had the misfortune to live in nineteenth-century Manchester, about a mile from the cotton exchange at the city's centre. The sky he woke to resembled no sky on Earth so much as some Victorian vision of hell: dark even at noon, with roiling plumes of black smoke. So foul was the Manchester air in 1848, at the height of the Industrial Revolution, that mothers, it was said, could barely make out the outlines of their children across the street.

I was driven to Edleston's former address by Professor Laurence Cook of the University of Manchester, who traced Edleston's home in an old city directory. He described the way the area looked in 1966 when he first moved there in terms reminiscent of the great industrial era: 'We'd take our daughter out in her pram and she'd be covered with black soot. People showed their housekeeping pride by wiping the edges of each step; they wouldn't even try to clean the whole thing.'

The novelist Elizabeth Gaskell, a contemporary of Edleston's, described a family's first sight of the city as they arrived by train: 'Quickly they were whirled over long, straight, hopeless streets of regularly built houses, all small and of brick. Here and there a great, oblong, many-windowed factory stood up, like a hen among her chickens, puffing out black "unparliamentary" smoke, and sufficiently accounting for the cloud which Margaret had taken to foretell rain.' The black smoke was 'unparliamentary' because even in Gaskell's time there were air-pollution laws on the books, though they were scarcely enforced. As much as 50 tons of industrial fallout settled over each square mile of the city every year. The trees, houses, and every other surface around Manchester and the other dark mill towns were coated with black soot. When summer rainstorms swept down over the Pennine hills, to the northeast, the pollutants ran down the tree branches in rivulets, destroying the lichens that grew on the bark. Soon all the tree trunks were stripped.

Rarities are prized by collectors, and the new black moth turned up in many collections in the second half of the nineteenth century. These pinned moths, together with the notebooks of Victorian collectors, enabled later scientists to reconstruct the spread of the mutant variety in the British Isles and Europe. Once established in a population, the black mutants often seemed to spread very rapidly. By the time Edleston wrote up his capture in a short, untitled note in the *Entomologist* magazine in 1864, the perplexing dark moths, which he called 'negroes', outnumbered the light typical moths in his garden. 'Last year I placed some virgin females in my garden in order to attract the males, and was not a little surprised to find that most of the visitors were the "negro" aberration,' he wrote. 'If this goes on for a few years the original type of *B. betularia* will be extinct in this locality.'

From Manchester the melanic peppered moths radiated throughout England during the second half of the nineteenth century. A *carbonaria* was caught in Cheshire in 1860, in Yorkshire in 1861, in Staffordshire in 1878, in London in 1897. By the 1890s, the moth population had shifted from light to dark in forests near industrial centres all over the country. A collector named W.F. Kirby noted in 1882 that 'the "pepper and salt moth" [an old name sometimes used for the species] has an almost black variety . . . which is not very uncommon in many parts of England.' By 1895 the *carbonaria* form had reportedly reached a frequency of 98 per cent in Manchester. There, as in several other places, it had all but replaced the typical form. It turned up in the grimier parts of the Continent as well. The first record of *carbonaria* in continental Europe was in 1867 in Breda, Holland, and one turned up in Hanover, Germany, in 1884. A bit later, melanic moths began showing up in the United States.

This dramatic change was observed by scientists, who proffered various explanations, all hampered by the rudimentary state of knowledge at the time. Although Gregor Mendel had published a paper on the laws of inheritance in 1862, it languished in obscurity until the turn of the century, so the origin of mutations as well as the mechanisms whereby offspring resembled parents were only dimly understood. Charles Darwin's bold new law of Natural Selection, which would in time become inextricably intertwined with the fate of the strange black moths, was still in its infancy, and much disputed.

In 1848, when Edleston first caught his moth, Darwin had not yet written his scientifically revolutionary work, *On the Origin of Species by Means of Natural Selection*. By 1864, when Edleston wrote up his discovery, Darwin's theory had been prominently launched, and he was one of the most famous people in Europe, but his ideas seem to have left no trace on Edleston. The entomologists who pondered the black moths during the latter half of the nineteenth century were inclined toward other explanations. Melanism, one William Tugwell speculated in 1877, was caused by 'the powerful impression of surrounding objects on the female during the all important period of life, viz. that of propagation, coupled with an instinctive provision for the protection of its future progeny'.[6]

Rearing eggs or caterpillars at low temperature and high

humidity produced melanic adults, some entomologists noted; therefore melanism might be the direct result of environmental pressures. Some entomologists floated speculations that were partly or wholly Lamarckian, following Jean-Baptiste Lamarck's theory that characteristics developed by an organism during its lifetime would be passed on to its offspring. (Thus the giraffe's long neck was believed to be the result of generations of giraffes stretching their necks to reach successively higher tree branches.) Even Darwin believed in the inheritance of acquired characteristics.

In a perfectly ordered universe, the dusky peppered moth caught by Edleston, then known as *Amphydasis betularia*, should have come to Darwin's attention and – just perhaps – triggered a revelatory moment. The moths were, in a sense, waiting for Darwin's theory.

Charles Darwin was born on 12 February 1809 in Shrewsbury, England, the fifth of six children. From an early age he was bored by school, and all his passions were directed to the outdoors, to fishing, collecting, hunting, and reading nature books. 'You care for nothing but shooting and dogs and rat-catching,' his worried father upbraided him, and it was true that he seemed destined for nothing more than country-squire obscurity. In his older brother's footsteps, he had been sent to the University of Edinburgh to study medicine, but he proved too squeamish, if not inept, and in 1828 was packed off to Cambridge to study theology instead. This was not entirely inappropriate: vicars in those days were apt to be natural historians, since a minute examination of nature was supposed to reveal the handiwork of God. In 1831 he was rather at loose ends when he was offered a position on HMS *Beagle* as a gentleman companion (but not as the official naturalist) to the captain, Robert Fitzroy, who had a commission to survey the coasts of South America.

There was little more *de rigueur* for a young naturalist in those days than a perilous sea voyage that lasted longer than a university education. In the heyday of the British Empire an insatiable Victorian curiosity about and appetite for exotic flora and fauna unleashed naturalists and collectors upon the most remote corners of the Earth. It was as if the British genteel class longed to swallow the globe whole, with its rare orchids, tropical butterflies, colourful natives and muddy, crocodile-infested rivers. The Victorians had a passion for order, an overarching, majestical, cosmic order, so when the specimens flowed back from Malaysia or Tierra del Fuego into museums and private collections, scientists were put to work arranging, cataloguing and naming them. What made all this possible, ultimately, was the wealth generated by the very factories belching the soot that may or may not have made melanic moths a fact of life in Britain.

The conventional version of Darwin's life pictures him having his transcendent insight while he was in the Galapagos Islands, but that is not quite accurate. Only after his return to England in 1836 did he became an evolutionist (although Darwin himself did not use that term; he called his theory 'descent with modification'). He had begun the voyage in 1831 believing, like his contemporaries, in the fixity of species. Every type of organism was a separate and immutable creation in the great chain of being, or *scala naturae*; each species was a 'thought of God', in the words of the eminent Louis Agassiz of Harvard, and its splendid adaptations to its habitat were signs of His loving care. Darwin's voyage changed him, however, as voyages are apt to do. As his relationship wore thin with the opinionated, Bible-quoting Captain Fitzroy, who had expected the voyage to substantiate the Book of Genesis and with whom he was forced to dine à *deux* every night, Darwin's isolation must have been extreme. He seized every opportunity to go collecting in remote places: volcanic and coral islands, the pampas of Patagonia, the Brazilian rainforests, Tierra del Fuego, the Andes. When he sailed back to England five years later it was with a shipload of specimens and fossils.

Upon his return he began sorting his collections and sending them off to specialists. The ornithologist John Gould of the Zoological Society of London was particularly enthusiastic about a series of peculiar ground finches that seemed to constitute a new group containing fourteen species apparently confined to the Galapagos Islands. Gould also pronounced the verdict that the three mockingbirds Darwin had bagged on three different Galapagos islands were three distinct species, and this, in Darwin's mind, was the judgement that undermined 'the stability of Species'. This moment in London, then, and not a Eureka experience in the distant Galapagos, was when Darwin became convinced that a new species could arise by the gradual accumulation of adaptations to new and different environments.[7]

As he sorted his collection, the evidence unfolded in species after species from the Galapagos: giant tortoises, iguanas, cactus trees, and finches that resembled their mainland cousins but were nonetheless distinct. His South American fossils, too, seemed to tell a story of organisms gradually evolving over millennia. By 1837, Darwin was convinced that all organisms on Earth were descended from a common ancestor and that new species had evolved gradually. However, he still hadn't hit on a mechanism. What exactly could have propelled life forms to change and develop as they did, instead of remaining static?

For a year and a half he produced and rejected one theory after another, until, on 28 September 1839, he arrived at his central illumination. It was the principle he would come to call

Natural Selection, and its catalyst was Thomas Malthus's 1798 essay on human populations. Darwin scholars believe they have isolated the particular sentence in Malthus that struck the chord: 'It may safely be pronounced, therefore, that the population, when unchecked goes on doubling itself every twenty-five years, or increases in a geometrical ratio.' If unhindered by war, famine, or disease, the population would quickly outgrow its means of subsistence. Since more individuals are produced than can possibly survive, Darwin reasoned that among all organisms ceaseless competition is the name of the game, and small differences assume life-and-death importance. Birds with a certain shape of beak, the 'swiftest of slimmest grey wolves', or bees with some minute difference 'in curvature or length of proboscis, far too slight to be appreciated by us' will win the race for existence.

Owing to this struggle, variations, however slight . . . if they be in any way profitable to the individuals of a species . . . will tend to the preservation of such individuals, and will generally be inherited by the offspring . . . I have called this principle, by which each variation, if useful, is preserved, by the term Natural Selection.

Although Darwin and his contemporaries did not yet know how inheritance worked, they knew that offspring resembled their parents. As individuals with a slight advantage – such as a better-designed beak, sharper claws, greater muscle strength, disease resistance, or camouflage – were more likely to survive and leave offspring (endowed with the same characteristics), while those with unfavourable traits were more likely to perish, Natural Selection would have a gradual, cumulative effect, shifting a species toward ever-greater adaptation. The philosopher Herbert Spencer (1820–1903), whose ten-volume work *Synthetic Philosophy* played a monumental role in promoting the idea of evolutionary progress, would call this principle the Survival of the Fittest, a term that Darwin embraced with relish.

Darwin knew of no specific examples from nature to support Natural Selection but rather argued his case by analogy. Having spent many congenial hours with pigeon-fanciers and country squires obsessed with the breeding of horses, cattle and dogs, he theorized that nature acts like the human breeder who, by selecting for certain traits, produces optimum breeds, or a wide range of very different breeds from dachshunds to German Shepherds. He even took up pigeon breeding himself and made the rounds of pigeon and poultry shows. At times his Natural Selection resembled a conscious agency tirelessly toiling over the perfection of creatures. 'Natural selection is daily and hourly scrutinizing, throughout the world, the slightest variations; rejecting those that are bad, preserving and adding up all that are good; silently and insensibly working ... at the improvement of each organic being in relation to its organic and inorganic life.'[8] Natural Selection would be even more powerful than artificial selection, he believed, for 'Man can act only on external and visible characteristics: nature cares nothing for appearances, except so far as they may be useful to any being. She can act on every internal organ, on every shade of constitutional difference, on the whole machinery of life.'[9]

The single example drawn from nature was the phenomenon of insect mimicry reported by the naturalist Henry Walter Bates in the course of his explorations of the Amazon with Alfred Russel Wallace between 1845 and 1859. Bates noted that some insects disguise themselves as natural objects, such as dry leaves or clods of dirt. In one form of crypsis, now known as Batesian mimicry, an edible insect has an appearance that copies the bright 'warning' coloration of a poisonous or unpalatable insect and thereby fools its predators. Darwin thought selection was the most plausible explanation for this phenomenon – a theory that would be debated fiercely decades after his death.

In fact, Darwin identified two distinct forms of selection in the *Origin*. The term *natural selection* is reserved for any trait that enhances survival, such as superior adaptation to climate, increased resistance to disease, or a greater ability to evade predators. But Darwin also saw that an individual who did *not* possess superior survival equipment might triumph simply by being better at reproduction. 'Sexual selection' was epitomized for Darwin in the flamboyant plumage of certain birds, the antlers of stags, and other attributes that could influence mating and reproductive success.

When the great comparative anatomist Thomas Henry Huxley, who became Darwin's greatest advocate, heard Darwin's ideas he is said to have exclaimed: 'How stupid of me not to have thought of that!' Here was a grand unifying principle linking all life forms from protozoa to emperors, and its essence was so simple, so transparent, that even a child could grasp it. There were only a couple of basic tenets: that there is heritable variation, and that there are differences in survival and reproduction among the variants. Some organisms survive and leave more offspring than competing organisms, and the winners' genes are passed on preferentially to future generations.

Darwin conceived his theory in 1839 but did not publish for another twenty years, becoming sidetracked for eight years on a monograph on barnacles. He had intended to produce an enormous tome about his evolutionary theory, and doubtless would have gone on writing and rewriting for many more years if A.R. Wallace had not independently developed essentially the same theory based on his observations in the Malay peninsula and written Darwin to ask his opinion. Obviously, Darwin could afford to procrastinate no longer, and on 1 July 1858 four papers by Darwin and Wallace were read before a meeting of the Linnean Society of London.

In the next year, on 24 November 1859, *On the Origin of Species by Means of Natural Selection, or the Preservation of Favoured Races in the Struggle for Life* – essentially an abstract of Darwin's unfinished tome – was published, and the 1,250 copies of the first edition were snapped up by booksellers on the first day. The publication quickly came to the forefront of Victorian consciousness, the disagreements it stimulated crystallized in the famous debate in 1860 between Samuel ('Soapy Sam') Wilberforce, the bishop of Oxford, who represented the creationist viewpoint, and Thomas Henry Huxley, who appointed himself 'Darwin's bulldog'. Huxley, an atheist, had launched his career as a naturalist on board a ship in the Pacific and then become a brilliant comparative anatomist, and he quickly embraced Darwin's ideas; it was he who coined the term 'Darwinism'. At the debate, held at Oxford's University Museum, it was reported that a few ladies fainted at some of the heretical ideas expressed in the room. According to legend, Wilberforce asked Huxley if it was through his grandmother or his grandfather that he claimed descent from a monkey, to which Huxley reportedly responded in the following vein: 'If the question put to me is "Would I like to have a miserable ape for a grandfather, or a man highly endowed by nature and possessed of great means and influence, and yet who employs these faculties and that influence for the mere purpose of introducing ridicule into a grave scientific discussion" – I unhesitatingly affirm my preference for the ape.' After this, undergraduates reportedly leapt onto their chair seats to cheer, and poor Captain Fitzroy, who was in the audience to cheer on the theological literalists, could be observed having a nervous breakdown, stomping around the room holding a Bible over his head. The details may have acquired mythical colour, but the story has endured. There is no doubt that Huxley was a phrasemaker; on the subject of Darwinism he also quipped: 'Extinguished theologians lie about the cradle of every science as the strangled snakes beside that of Hercules.'

Like the Bible and the works of Shakespeare, the *Origin* is one of the world's key texts, endlessly cited and subject to the most varied interpretations. Scholars have tirelessly disputed Darwin's faith in God, as well as his beliefs about whether evolution is purposeful or totally random. The renowned Harvard paleontologist George Gaylord Simpson would write, a century later, that evolution 'achieves the aspect of purpose without . . . a purposer, and has produced a vast plan without a planner.'[10] Certainly a review of the past several hundred million years reveals a flow in one direction, namely from simplicity to complexity. How, then, could such apparent order and purpose have come about by trial and error? In his autobiography Darwin confessed himself overwhelmed by 'the extreme difficulty or rather impossibility of conceiving of this immense and wonderful universe . . . as a result of blind chance. I felt compelled to look for a First Cause.' Later, he wrote: 'My theology is in a simple muddle. I cannot look at the universe as a result of blind chance, yet I see no evidence of beneficent design in the details.'

That death and suffering loomed large in the script was an unhappy fact of life. Death provides the fuel for Natural Selection, for the more (selective) deaths occur, the more powerful the selection. The losers, the less fit, must be swiftly eliminated. In Darwin's universe, 'nature red in tooth and claw', in Tennyson's phrase, is exposed in its harshness, and this was reason enough for some sentimental Victorians to reject it. 'We behold the face of nature bright with gladness,' Darwin wrote plangently in the 'Struggle for Existence' chapter, '. . . we often do not see, or we forget that the birds which are idly singing round us mostly live on insects or seeds, and thus are constantly destroying life; or we forget how largely those songsters or their eggs, or their nestlings, are destroyed by birds or beasts of prey; we do not always bear in mind that though food may be superabundant, it is not so at all seasons of each recurring year . . .'[11] The only consolation this man who was too tender-hearted for medical school could proffer was that, according to the missionary/explorer David Livingstone, who had once been rescued from the jaws of a lion, being eaten by a big cat was not really such a bad way to go.

It is widely agreed that Darwin embarked on the *Beagle* a conventional Christian who embraced the literal truth of the Bible and was steeped in the tradition of natural theology epitomized by the clergyman William Paley (1743–1805). In the celebrated opening paragraph of his influential book *Natural Theology*, Paley asks the reader to suppose that he had stumbled on a watch for the first time:

We see a cylindrical box containing a coiled elastic spring, which, by its endeavour to relax itself, turns round the box. We next observe a flexible chain . . . We find a series of wheels . . . We take notice that the wheels are made of brass, in order to keep them from rust . . . that over the face of the watch there is placed a glass . . . the inference we think is inevitable, that the watch must have had a maker – that there must have existed, at some time at some place or other, an artificer or artificers who formed it for the purpose . . .[12]

In a similar manner, Paley argued, many biological systems such as muscles, bones and mammary glands are exquisitely designed and could not function if one or several components were lacking. Their intricate complexity argues for a divine designer. This is a classic statement of the theological 'argument from design', which the present-day Oxford Darwinist Richard Dawkins set out to refute in his book *The Blind Watchmaker*, addressing Paley explicitly. But if Natural Selection could provide a mechanistic explanation for the intricate designs of organisms, the argument from design could be retired, as Darwin realized. 'We can no longer argue that, for instance, the beautiful hinge of a bivalve shell must have been made by an intelligent being, like the hinge of a door by man,' he wrote in his autobiography. 'There seems to be no more design in the variability of organic beings and in the action of natural selection, than in the course which the wind blows.'

The natural world that Darwin observed so meticulously flatly contradicted the biblical version of creation, for 'creation simply could not explain the fossil record, nor the hierarchy in types of organisms that had been proposed by the taxonomist Carl Linnaeus, nor many of the other findings of science', as the celebrated ornithologist Ernst Mayr wrote. 'Yet almost all of Darwin's peers still believed in some form of creation, and many of Darwin's contemporaries accepted Bishop Ussher's calculation that creation had occurred as recently as 4004 B.C.'[13] Darwin has been called a deist, an agnostic and an atheist, and each of these views can find support in his writings. Some scholars insist that he preserved the outward forms of religion and kept his anguished doubts to himself in deference to his devout and much-loved wife, Emma, who was his model for the proper Victorian reader. Consider the famous poetic and strangely moving 'tangled bank' passage in the *Origin*, the last words in the book:

It is interesting to contemplate a tangled bank, clothed with many plants of many kinds, with birds singing on the bushes, with various insects flitting about, and with worms crawling through the damp earth and to reflect that these elaborately constucted forms, so different from each other, and dependent upon each other in so complex a manner, have all been produced by the laws acting around us. These Laws . . . being Growth and Reproduction; Inheritance, which is almost implied by reproduction; Variability . . .; a Ratio of Increase so high as to lead to a Struggle for Life, and as a consequence to Natural Selection . . . Thus from the war of Nature, from famine and death, the most exalted object which we are capable of conceiving, namely the production of higher animals, directly follows. There is a grandeur in this view of life, with its several powers, having been originally breathed by the Creator into a few forms or into one; and that, whilst this planet has gone cycling on according to the fixed law of gravity, from so simple a beginning endless forms most beautiful and most wonderful have been, and are being evolved.

Here is Darwinism at a glance: the whole pageant of life emerging from simple beginnings and owing its existence to the operation of a handful of laws as impersonal as the law of gravitation. But what is this 'breathed by the Creator'? Is Darwin saying that God created the first life forms and Natural Selection did the rest? Even more interesting, the phrase 'breathed by the Creator' appears in some editions and not in others. Darwin went to his grave in 1882 without saying definitively whether or not he believed that God was dead, leaving it to his successors to sort out the theology.

Darwin also died without resolving the sticky points of his theory, of which he was painfully conscious and with which he wrestled mightily. One stumbling block concerned the absence or rarity of transitional forms in the fossil record – where are the 'missing links'? A second was the 'unbridgeable gap'. Darwin blithely wrote that he saw no difficulty in theory in a bear becoming a whale, but how does a carnivorous land animal develop into an aquatic animal, given that the transitional form would seem to have a compromised fitness? A third, related difficulty was posed by organs of extreme perfection. How does a blind force like natural selection create organs as intricate as the eye? Half an eye may be no better than no eye at all. A further problem, the mystery around the source of variation, along with Darwin's erroneous idea of 'blended' inheritance, would be erased when Mendelism became fused with Darwinism in the 1920s. The other three problems are still being debated today.

The world had been prepared for evolutionary theory by the earlier evolutionary ideas of Lamarck and others. A century earlier the Swedish naturalist Karl von Linné (1707–78), known as Linnaeus, had discerned a hierarchical pattern of natural groupings within the plant and animal kingdoms in his *Systema Naturae*. After Darwin, the Linnaean hierarchy suddenly became quite logical, for it now appeared that each higher taxon, or distinct group of organisms, consisted of the descendants of a still more remote ancestor, and biogeography, the study of how and why plants and animals came to be where they are, seemed to confirm evolutionary theory. By the 1880s the majority of biologists in Britain had accepted evolution. Natural Selection, however, remained controversial. The randomness it embodied, and the relentless destruction of the less fit, ran foul of deeply rooted ideas of beneficent design. Another conceptual difficulty was that Victorians were accustomed to thinking of organisms in terms of 'types', and could not grasp that Natural Selection involved changes in populations, not individuals, for individuals themselves do not evolve.

Darwin believed that minor individual differences provided the variation upon which Natural Selection operated and that the process of evolution was extremely gradual. 'The mind cannot grasp the full meaning of the term of even a million years,' he wrote; 'it cannot add up and perceive the full effects of many slight variations, accumulated during an almost infinite number of generations . . .'[14] Many of his contemporaries disagreed, and looked to large, freakish mutations, known as 'sports', as the stuff of evolution. In these saltationist theories (from the Latin *saltum*, or 'jump') the mechanism of evolution was not an accretion of small, imperceptible differences but a sudden transmutation. Huxley himself advised Darwin: 'You have loaded yourself with an unnecessary difficulty in adopting *Natura non facit saltum* ["Nature does not make jumps"] so unreservedly.'

At the time of Darwin's death Natural Selection (and gradualism) was becoming increasingly unfashionable. In addition to ever-popular saltationist theories, other alternatives were floating around. The inheritance of acquired characteristics, a concept to which Darwin subscribed, was thriving. In its Lamarckian version, changes were said to result from the effects of use or disuse. Another theory, called orthogenesis, of which Henri Bergson and Teilhard de Chardin were later adherents, posited that some inner, vitalist force impels organisms to evolve toward greater complexity.

After Darwin's death, it is said, an unopened package was found among his effects containing a paper by Gregor Mendel, a fact that has always provided a tantalizing 'what if?' for Darwin scholars: What if Darwin had opened his mail and perceived that here in his hands was the solution to the puzzle of inheritance? Would evolutionary science have been spared fifty years of blind alleys and detours? In fact, Darwin may have just missed intuiting the machinery of heredity on his own. 'Once, Darwin almost got it right,' according to the British geneticist Steve Jones. 'He noticed that the young from a cross between two different stocks of pigeons were uniform, but that when these mongrels were crossed for several generations then hardly two were alike. Mendelism is, we now know, at work, as the stocks differ in several genes that later come together in many ways.'[15]

To this we might add a secondary and more modest 'what if?' What if Darwin had fancied Lepidoptera instead of pigeons? What if he'd been in the habit of reading entomology journals and had come across reports of the strange black moths in the industrial regions? Would his thoughts have strayed to Bates's inquiries into insect mimicry, especially those cases of impeccable camouflage that he felt demonstrated the action of Natural Selection? Would he have seized on the melanic form of *Biston betularia* as a living illustration of evolution?

When the first melanic peppered moth turned up in Britain, Darwin was still absorbed in his 'beloved barnacles'. At the time of his death, no black peppered moths had yet been sighted in his rural county of Kent, though by the middle of the next century there would be melanics aplenty there. Whether he ever heard that some populations of British moths were turning black we'll never know.

The first person to connect peppered moths with Darwin's theory was J.W. Tutt, the editor of *The Entomologist's Record and Journal of Variation* and the foremost lepidopterist of the Victorian age. Initially, Tutt had ascribed melanism to humidity and other environmental factors; by 1896, he had come around to Natural Selection. Like most species, this night-flying moth passes the day motionless on the trunks and branches of deciduous trees and other surfaces. As long as it is well camouflaged against pale and lichened surfaces, Tutt proposed, the typical form has some protection against birds. However, as the trees and all other surfaces become soot-coated, the tables turn:

But some of these peppered moths have more black about them than others, and you can easily understand that the blacker they are the nearer they will be to the colour of the tree trunk [in industrial regions], and the greater will become the difficulty of detecting them. So it really is; the paler ones the birds eat, the darker ones escape. But then if the parents are the darkest of their race, the children will tend to be like them, but inasmuch as the search by birds becomes keener, only the very blackest will be likely to escape. Year after year it has gone on, and selection has been carried to such an extent by nature that no real black and white peppered moths are found in these districts but only the black kind. This blackening we call melanism.

The typical moths thrive in rural, unspoiled forests because their camouflage is effective against a background of lichens growing on the tree trunks. Their avian predators cannot see them. In the industrial areas, pollutants kill the lichens and darken the trunks. The melanics are better adapted to this new environment, being camouflaged against dark surfaces. They now have the selective advantage. It was Darwinian Natural Selection in a nutshell, just as the nineteenth century was drawing to a close.

But in 1896 nobody believed it. It wasn't only because many people were confused about Natural Selection, or because, before the rediscovery of Mendel's work, there was no known mechanism to explain how the moths had become dark in the first place. Tutt's Darwinian view was vigorously challenged, according to Michael Majerus of Cambridge, 'primarily because entomologists and ornithologists did not regard birds as major predators of cryptic day-resting moths'.[16] In short, Tutt's hypothesis was disbelieved because no one had observed birds eating peppered moths.

3

Natural selection reduced to arithmetic

The year 1900, greeted by some with apocalyptic dread, was the penultimate year of Queen Victoria's reign, which had lasted so long that no one under seventy could remember any other monarch. At her diamond jubilee three years earlier, the *Times* perorated: 'No State or no Monarch known to history has ever rejoiced in such homage as our colonies will pay to our Queen.' Likewise, scientists devoted to Darwin's theory were paying homage to the mysterious moths, black-clad, like the dowager queen, that had appeared like harbingers among them. In 1900, the Evolution Committee of the Royal Society of London singled out industrial melanism as a phenomenon of urgent scientific importance, for it was evident that something unprecedented was occurring. Unfortunately the entomological records proved too patchy to provide reliable data.

Another event occurred in 1900, which would change the fate of the peppered moth. Riding in the polished mahogany depths of his carriage on the Great Eastern Railway between Cambridge and London on 8 May 1900, William Bateson, a forty-year-old don of St John's College, Cambridge, became immersed in an old article from an Austrian journal written by an obscure monk, Gregor Mendel, abbot of the Augustinian monastery at Brunn (now Brno) in Moravia. The article, dating from 1865, was nearly as old as Bateson himself, but the don happened to be reading it because it had just been cited separately by three scientists in 1900. By the time he disembarked into the clamour of London's Liverpool Street Station, a perspicacious observer might have noticed he had the glow of the newly converted. He discarded the speech he had planned to deliver to the Royal Horticultural Society and instead began broadcasting the message of Mendelism. 'An exact determination of the laws of heredity will probably work more change in man's outlook on the world, and in his power of nature, than any other advance in natural knowledge that can be foreseen,' he told the audience. 'There is no doubt that these laws can be determined.'

In a series of elegant breeding experiments with the pea plant, Gregor Mendel had chosen easily identifiable 'characters' such as tallness and shortness, yellow or green pea colour. When he bred a tall plant with a tall plant, the offspring were all tall. When he crossed a short plant with a short plant, the offspring were all short. When he crossed a tall plant with a short plant, the first group of offspring were all tall, but when he crossed certain tall plants from that generation with each other, he always got, in the subsequent generation, the interesting ratio of three tall plants to one short plant. The fact that the short plants all 'bred true' – produced more short plants – when crossed with other short plants meant that the trait of shortness had been masked by the tallness character in one generation but reappeared unchanged in the next.

Mendel's ratios were always the same in the second generation: three tall plants to one short plant, three plants producing yellow peas to one plant with green peas. Mendel proposed that every organism had two 'factors' for a given trait, one from the mother and the other from the father. Both factors could be alike in a given organism, or they could be unlike. Body cells contain a pair of factors for each character, while each germ cell has only one, which is passed on to the offspring. Mendel's 'factors' would later become known as allelomorphs or alleles. In offspring the alleles can combine as two yellow alleles, a yellow and a green, or as two greens. A pea plant with one yellow allele and one green allele will have yellow peas because the yellow allele is 'dominant', concealing its 'recessive' partner. Yet each hybrid contains a copy of the green allele and a copy of the yellow. Simple arithmetic shows that when the hybrid plants are intercrossed, the next generation will consist of one yellow–yellow, two yellow–greens, and one green–green. The yellow–yellow and the yellow–greens will appear yellow, and the green–green will appear green: a ratio of three to one.

Darwin had subscribed to the idea of 'blended inheritance', according to which the characteristics of the parents are mixed in the offspring. This notion caused problems for his theory, for it was evident that blending would homogenize traits and eventually there would be no black or white rabbits in the world, only grey ones. Mendel's laws solved this difficulty. From his experiments Mendel deduced that, in the formation of germ cells, the two factors for any trait always separate from each other and end up in different eggs or sperm. The inherited material came in the form of distinct, particulate units, which did not become 'blended' in the organism; thus traits apparently lost in one generation could reappear in another. 'Mendel's theory of heredity was the perfect complement to Darwin's theory of natural selection,' observes the historian of science William Provine of Cornell University. 'Mendelian characters could be very small and were not blended away by crossing. Furthermore, Mendelian recombination provided new variability for selection.'[17]

Despite this revolutionary new information, it would take twenty years for the Darwinians to appreciate what Mendelism had to offer. In 1900 Darwinian theory remained incomplete and rather muddled. Systematists and taxonomists who embraced Darwin were engaged in tracing wholly speculative family trees for different organisms. Darwinian biometricians, who worked on the puzzle of heredity, had equations for calculating the resemblances of various traits between parents and offspring, but, like Darwin, they could not explain how variations could persist in a population without being 'swamped'. It was the old problem of the grey rabbits, which a particulate notion of heredity could have dispelled in an instant, but to the biometricians the news from Brno was about as welcome as the plague.

A series of misunderstandings between the Darwinians and the Mendelians would keep them at odds for years. Just after the turn of the century, a Dutch plant-breeder, Hugo De Vries, was struck by the sight of two distinct species of evening primroses growing side by side in a field outside Amsterdam. He bred them and demonstrated experimentally that a new species could arise in a single jump. De Vries's book *The Mutation Theory* (1903) presented an attractive alternative to the (by this time) rather unpopular idea of Darwinian natural selection. Instead of waiting for the slow, exhausting accumulation of tiny variations, new species could originate in a single generation through the occurrence of large-scale variations, called 'mutations'. It later turned out that the evening primrose was the exception in its ability to mutate into a new species virtually overnight, but in the meantime the new Mendelians had become 'mutationists'. In their view, mutation pressure alone was the force that propelled evolution, with new species coming into being when sufficiently large mutational jumps occurred. There was little need now for natural selection. The Darwinians, for their part, turned their backs on Mendelism, thereby effectively becalming evolutionary theory for two decades, and both camps went their separate ways, speaking their incompatible languages.

The Mendelian 'factors' had begun as abstractions, but they were found to have a material existence. Unlike the speculative turn-of-the-century Darwinians, the Mendelians between 1910 and 1930 were developing a pragmatic, experimental, and quantitative approach to biology. In the legendary 'fly room' at Columbia University's Shermerhorn Hall, Thomas Hunt Morgan and his colleagues raised thousands of fruit flies (*Drosophila*) in milk bottles and developed methods for mapping the chromosomes, which were conveniently located within the cells of their huge salivary glands. Because a new fruit-fly generation occurs every ten to fourteen days, the data accumulated rapidly. Chromosome markers were developed and as successive generations were bred, traits such as eye colour and bristle number could be correlated with studies of chromosome structure. By breeding certain lines, the scientists could create almost any sort of fly they liked – flies with stunted wings or vermilion eyes or lots of bristles. They were also witnessing mutations arising before their eyes. Morgan's student H.J. Muller found that he could produce mutations more frequently by raising the temperature in the milk bottles or by exposing the flies to the science-fiction-like radiation discovered by W.R. Röntgen in 1895. If you wanted more mutations, all you needed was an X-ray machine.

Genetic advances revealed that mutations – essentially copying errors in hereditary material – were the source of all the variation that was necessary for evolution. The 'point mutations' identified in the fly room, consisting of small changes in the DNA nucleotide sequence, were not the huge one-step 'mutations' identified by De Vries. The use of the same word for different phenomena caused confusion, and was one reason that many Darwinians abhorred Mendelism for as long as they did. The schism between the geneticist/experimentalists and the naturalists deepened in the first decades of the twentieth century, the former despising the latter as unscientific and the latter brushing off the events of the fly room as entirely artificial.

The man who would unite these two feuding worlds had been a schoolboy when William Bateson read Mendel's paper, a near-sighted ten-year-old British mathematical prodigy named Ronald Aylmer Fisher. Young Ronald's severe myopia doomed him to lenses the thickness of Coke-bottles, and his childhood was further marred by a 'constricted' emotional communication with his mother, who, according to Fisher's daughter and biographer Joan Fisher Box, passed on to her son her own 'inadequate emotional vocabulary'.[18] All his life Ronald would be emotionally insecure, curiously oblivious to the emotions of others, often abrasive, and semi-paranoid when his will was frustrated. His outbursts of temper, 'like the thunderbolts of Jove', brought everything around him to a standstill; once, in a fury against a laboratory assistant, he crushed the mouse he was holding in his hand and then exclaimed: 'Look what you've made me do!' There could be no doubt of his genius, however. To avoid eyestrain his doctor had forbidden reading by electric light, and as a pupil at Harrow he had to be tutored in mathematics without the aid of a pencil and paper or blackboard. Out of this disability came an uncanny ability to work out mathematical problems in his famously oversized head, and all his life he would leapfrog effortlessly over intermediate steps to a solution, leaving everyone else in the dark about how he had arrived there.

His father went bankrupt just before it was time for Ronald to go to university, so it was as a scholarship boy that he went up to Gonville and Caius College, Cambridge, in 1909, the centennial of Darwin's birth and the fiftieth anniversary of the publication of the *Origin of Species*. At an international gathering at Cambridge in July T.H. Huxley's widow, dignified in black bombazine and bonnet, and the even more ossified widow of Joseph Hooker, Darwin's botanical ally, were paraded out, and many people spoke wistfully of the so-called good old days. An undercurrent of sadness permeated the air, for in 1909 Darwinism in England had reached its nadir. These were the dark years that Julian Huxley, T.H. Huxley's grandson, then a student at Oxford, would call the 'eclipse of Darwinism', when competing theories such as orthogenesis, aristogenesis and Lamarckism were more fashionable than natural selection. For those who felt that the Darwinian jig was almost up, the gaps in Darwinian theory loomed large and many biologists were turning against it. Nonetheless, the 1909 proceedings were published as a boxed set and found their way into the hands of the young Ronald Fisher, for whom they were intellectual fodder just as he began his first year at university. At Cambridge Ron's social deficits and indifference to matters of dress did not deter him from forming friendships. He became part of a lively, rather avant-garde coterie called the We Threes, who spoke their own idiom, composed of phrases from *Thus Spake Zarathustra*, read Icelandic sagas and incorporated names and phrases from them in their talk. Ron was considered an intellectual giant, but uncouth and strangely naïve.

He took a degree in astronomy, and his interests were more

mathematical than biological (he would become one of the great statisticians of all time). However, all the force-lines of his destiny were aligning around him, for Cambridge was the Mecca for Mendelism in England, largely due to the presence of William Bateson, who occupied the Arthur Balfour Chair of Genetics. Bateson and a few others had initiated breeding experiments along Mendelian lines and were making fundamental discoveries in the new science. Some of these ideas found their way into Fisher's brain at the same time that he was imbibing the statistical theories of the Darwinian biometricians, notably the vehemently anti-Mendelian Karl Pearson. (For a time Pearson was Fisher's mentor, and their eventual break proved legendary for its bitterness.) The other passion in Fisher's life was eugenics, to which he was introduced by the man who would become a benefactor and adoptive father to him, Charles Darwin's son, Major Leonard Darwin. Perceiving Fisher's great gifts, Leonard Darwin, who was president of the Eugenics Education Society of London, took the lad under his wing, frequently finding him jobs and keeping him afloat financially during his lean years.

Fisher became fixated on the idea that in the civilized world the laws of natural selection were relaxed and the least fit humans were provided a safety net. Thus the genes of Englishmen were daily becoming slacker, in tandem, some might say, with the aging British empire's weakening grasp on its far-flung colonies. Lamenting the declining fertility rates of the British upper middle classes, which were synonymous in his mind with the 'fittest', Fisher concocted a scheme of government allowances to encourage the 'fit' (that is, doctors and lawyers as opposed to unskilled workmen) to have numerous children. These ideas, then considered 'progressive', were a familiar part of the late Edwardian zeitgeist, but Fisher was unique in actually taking his own advice to heart and trying earnestly to lead a 'eugenic life'. He married a beautiful woman, began a family that would expand to eight children, left academia and tried to make a living as a subsistence dairy farmer. His reasoning was that farming was the only trade for which having many children was an asset. In practice he was a disastrous farmer and he and his family had to be rescued by Leonard Darwin.

Fisher's contribution to evolutionary biology proved incalculable. With the two other founding fathers of population genetics, J.B.S. Haldane and Sewall Wright, he brought about the fusion of Mendelian genetics and evolutionary theory and forged the theorems that became the hard core of modern Darwinism. Out of this merger would come the famous Synthesis, as it was dubbed by Julian Sorell Huxley, Thomas Henry Huxley's zoologist grandson who, with the family gift for nomenclature, would coin such neologisms of the field as 'cline', 'clade' and 'ethology'. Up to this time, the key concept of 'fitness' was qualitative and a bit elusive. Darwin viewed it as a particular design feature or way of life – long legs on a horse, a well-designed proboscis on a bee – that would enable the organism to carry out the business of living and reproducing. R.A. Fisher was not interested in scrutinizing the organism's way of life, or its proboscis or antlers. The sole criterion of fitness became the number of offspring left by particular organisms. It boiled down, ultimately, to genes.

First, Fisher applied himself to proving that inheritance was Mendelian and that individual 'factors' survived intact rather than becoming blended. In 1918 he published a paper showing that Pearson's data on the 'correlation between relatives' in stature and other measurements matched the Mendelian model. Next he took on the Herculean task of applying numbers to evolution, quantifiying the evolutionary consequences of Mendelian heredity. Calculating the effects of selection on single genes, he came to the conclusion that natural selection was slow and deliberate. Evolutionary change, Fisher believed, could yield universal laws similar to the physical laws governing the behaviour of gases.

By 1930, as the doctrine of Fascism was taking hold on the Continent, Fisher had his theory all worked out and published it in the manifesto *The Genetical Theory of Natural Selection*.[19] 'In the future,' he wrote, 'the revolutionary effect of Mendelism will be seen to flow from the particulate character of the hereditary material. On this fact a rational theory of Natural Selection can be based, and it is, therefore, of enormous importance.' In contrast to the macromutationists, Fisher assumed that 'mutations cannot have the power to direct the course of evolution [for] most produce deformities, often lethal'. The mutations that fed new variations into the gene pool had to be of minor effect, and the driving engine of evolution must be natural selection. The book's centrepiece was the Fundamental Theorem of Natural Selection, which Fisher arrived at by combining Mendelism with certain concepts from the ecology of populations. He started with actuarial tables and ended up with a general statement on rates of change of fitness, defined mathematically. According to the theorem, a population is always increasing in fitness, but because the environment is constantly changing – 'deteriorating', actually – the population never reaches optimum fitness and thus is condemned to race forever toward an ever-receding goal. (This concept has been dubbed the Red Queen's dilemma, with a nod to *Through the Looking Glass*.)

To craft a workable mathematical model of natural selection, Fisher made a number of simplifying assumptions, such as treating genes as though they did not interact at all. In his view selection was a deterministic process, grinding away relentlessly on the variation in the gene pool. If a certain gene bestowed an advantage that resulted in enhanced reproduction, he could calculate how rapidly it would increase in frequency. He found that a selective advantage as small as 1 per cent would add up more quickly than one would suppose. 'If we speak of a selective advantage of one percent, with the meaning that animals bearing one gene have an expectation of offspring only one percent greater than those bearing its allelomorphs [the alternative versions], the selective advantage in question will be a very minute one ... Such a selective advantage would, however, greatly modify the genetic constitution of the species, not in 100,000 but in 100 generations.'

The second half of Fisher's book is devoted to his eugenics agenda. Eugenics later became tainted by its association with Nazi racial theories, but in the 1920s and 1930s it was a widespread enthusiasm associated with 'enlightened' causes. In a broad sweep Fisher surveyed the collapse of empires, the decline of British peerages, and such matters as the proper distribution of family allowances to people of superior beauty, intellect, health and talent, whose breeding ought to be encouraged. It didn't seem to have occurred to him that in the state of nature, lacking optometrists, he might not have been the fittest of the lot.

An important influence on Fisher was a book by an ideological foe, the Mendelian J.C. Punnett, Bateson's successor to the Balfour Chair of Genetics at Cambridge. As a 'mutationist', Punnett did not grant much scope to natural selection and ridiculed the Darwinians' sloppy habit of making wild armchair conjectures about the adaptive value of different traits. He did concede, however, that natural selection might play a minor role once a mutation had occurred, and in his influential 1915 book *Mimicry in Butterflies* he wrote:

The case of the peppered moth shows how swiftly a change may come over a species. It is not at all improbable that the establishing of a new variety at the expense of an older one in a relatively short space of time is continually going on ... much could be learned if some common forms were chosen for investigation ... Large numbers should be caught at stated intervals, large enough to give trustworthy data as to the proportions of the different forms that occurred in the population. Such a census ... if done thoroughly and over a number of years at regular intervals, might be expected to give us the necessary data for deciding ... whether there were definite grounds for supposing natural selection to be at work, and if so what was the rate at which it brought the change about.[20]

Even though Fisher would devote himself wholeheartedly, in a landmark 1927 paper, to *refuting* Punnett's theory of mimicry, Punnett's novel idea of monitoring a particular genetic trait in a natural population in order to gauge the rate of natural selection reads ironically like a mission statement of the future work of Fisher and his protégé E.B. Ford, who would become H.B.D. Kettlewell's boss.

The Synthesis had indeed arrived. 'In a way, the synthesis was nothing but a confirmation of Darwin's original theory,' Ernst Mayr would write in 1974,[21] 'even though Darwin had published prior to the development of genetics and cytology [the study of the cell], and had been forced to treat the origin of variation as a black box. His basic theory, that evolutionary change is due to the combination of variation and selection, was, however, completely sound and is daily confirmed by every evolutionist.' The men whose theorems launched the Synthesis were a warring triumvirate; far from collaborating, they usually shot at each other with their competing equations. Fisher, politically conservative and a god-fearing member of the Church of England, was ill at ease with J.B.S. Haldane, who was a boisterous iconoclast and a Marxist, and with the American Sewall Wright he came to the academic equivalent of blows. Wright, an animal breeder for the US Department of Agriculture, had worked under the great W.E. Castle of Harvard. A self-taught mathematician, he quantified changes in the genetic composition of rat populations resulting from mutation and natural selection and he also investigated another factor: the effect of chance in small populations. This phenomenon was called 'random genetic drift', and for Fisher, who liked his evolutionary biology deterministic, it was a *mauvais génie* against which he would fight to the death.

The third of the great population geneticists, John Burdon Sanderson Haldane, was the brilliant, idiosyncratic, polymath son of the prominent Oxford physiologist John Sanderson Haldane. As a teenager he took up the breeding of guinea pigs with his sister Naomi, and stumbled on one of the most important phenomena of genetics: 'genetic linkage', the fact that some genes are linked to nearby genes on the same chromosome and are passed on together. The Great War interrupted his studies. Haldane was the only one of the original population geneticists to have experienced the trenches. Strangely, he rather savoured the experience and later, in his signature booming voice, would regale dinner companions with graphic and enthusiastic descriptions of the effects of explosives on the human anatomy. While he had taken a first-class degree in Greats (classics) at Oxford, he was also well versed in mathematics and physiology and for a time taught physiology at Oxford. During that period Julian Huxley recorded that 'Jack' had a habit of dropping in to his New College rooms around teatime and devouring 'plates of biscuits, protesting that he couldn't eat a crumb, while reciting Shelley and Milton and any other poet you chose by the yard'.[22] When he tired of Miltonic verse, he was apt to recite Homer in Greek at the drop of a hat. Julian Huxley's brother Aldous would use Haldane as his model for both the absent-minded, kidney-preoccupied physiologist Shearwater[23] and the degenerate Coleman in his 1923 novel *Antic Hay*, or so everyone said.

Despite his peculiar enthusiasms, Haldane's mind was a superb mathematical tool, and in a series of papers from 1924 to 1931 he constructed theoretical models of populations, plotting the factors that influenced the fitness values and frequencies of particular genotypes. He studied some specific conditions that Fisher had ignored for the sake of simplicity, but he too emphasized selection applied to single genes, and his models were fairly remote from the realities of woodland and meadow. He loved to shock established institutions, and was subject to crazes. In 1924 he gave a paper before the Heretics Society in Cambridge mentioning the possibility of ectogenesis (birth outside the body), which made him the butt of donnish jokes around Oxford and so mortified his venerable father, Dr Haldane the elder, that his mother was moved to ask Huxley to persuade people to stop poking fun at it. But it was another paper by Haldane, also published in 1924, that every evolutionary scientist knows, and it concerned the peppered moth.

Haldane had an inkling that evolution might work much faster than Fisher thought. When he needed a real-world example, it was natural to choose the celebrated *Amphidasys betularia*, as it was then called, whose mutant black form had evidently undergone such a spectacular increase in the latter half of the nineteenth century. Haldane reckoned that if the first melanic had been sighted in Manchester in 1848, it must have had a frequency lower than 1 per cent of the population, and by 1901 it had all but replaced the typical form in the region. Using these figures, he showed that in this industrial area the melanic form of the peppered moth would have had a selective advantage of 50 per cent – in other words, a 50 per cent greater production of offspring. This paper was a bolt out of the blue. If such intense selection pressures existed in nature, no one could doubt that natural selection was a powerful force, propelling evolution with no help from other mechanisms such as macromutations. Of course, Haldane never counted any peppered moths himself, or observed what happened to them; he simply fed the numbers into his model and out popped a figure of the rate of evolution. Thirty years would pass before anyone went out and tested Haldane's theory. First, it was necessary to develop the tools.

It was in the fertile Synthesis years of the 1930s and 1940s that the scriptures of modern Darwinism were produced at a breathless pace. In addition to the seminal works of Fisher, Haldane and Wright, there were influential tomes by the biologist Julian Huxley, the geneticist Theodosius Dobzhansky, the botanist Ledyard Stebbins, the paleontologist George Gaylord Simpson, the systematist Ernst Mayr, among many others. By the end of the 1940s, the Synthesis was complete, and neo-Darwinism, as the more scientific reincarnation of the original Darwinism was called, was launched. The field had defined its core principles and banished the various heresies that had bedevilled it: orthogenesis and aristogenesis, the inheritance of acquired characteristics, saltationism, and the rest. Natural selection (acting on mutations) was the only accepted mechanism of evolutionary change. What emerged was a coherent evolutionary theory with a lingua franca and a set of principles that brought a number of formerly disparate sciences, including taxonomy, systematics, paleontology, genetics, botany, zoology and ecology, under one roof.

The terms and concepts that would form the idiom of the new priesthood of population geneticists were coined during these years. The crucial distinction between genotype and phenotype was made. The phenotype is the organism's appearance and physical and behavioural properties; the genotype, the hereditary 'factors' that determine the phenotype. Evolution was said to consist of the addition or subtraction of genes from the 'gene pool', the sum total of all the genotypes in a breeding population. For example, some human beings can roll their tongues, while others cannot, and this ability is genetically determined. In this case, the description of the gene pool would specify the frequency of tongue-rolling and non-tongue-rolling alleles in the population.

By the 1940s the basic dramas of natural selection were reduced to arithmetic. Evolution was redefined as a change in gene frequencies, and natural selection was said to take place when the carriers of one allele (call it 'A_1') are more successful in reproducing than the carriers of an alternative allele (A_2). This reproductive advantage must take place consistently and systematically over successive generations. The 'survival of the fittest' could now be quantified, though of course it is not really the fittest that survive but simply the *fitter*, or even the adequately fit. The carriers of a certain allele may survive more than the carriers of a competing allele, or they may be more fertile or reproduce earlier, or they may have flowers more alluring to the insects that pollinate them. Any advantage or a combination of them may give one allele a higher *fitness* than another. Fitness is a relative measure.

For all his ardour, T.H. Huxley could not conceive of Darwinism as an experimental science: he noted that the author of the *Origin* 'does not so much prove that natural selection does occur, as that it must occur, but in fact no other sort of demonstration is obtainable.' Darwin himself had confided, in a letter to the botanist George Bentham: 'In fact, the belief in Natural Selection must be grounded entirely on general considerations [theory] . . . When we descend to the details, we can't prove that any one species has changed . . . nor can we prove that the supposed changes are beneficial, which is the groundwork of the theory. Nor can we explain why some species have changed and others have not.'[24] Indeed, the first attempts to demonstrate natural selection with experimental breeding failed miserably, proving just how hard selective breeding work can be. You need proven genetic variation. You need characters that are reasonably easy to measure. You need experimental controls to avoid contamination of results. The first really successful experiments were performed by William Ernest Castle of Harvard in the second decade of the twentieth century.

Castle worked with a breed of piebald rats known as hooded rats. The standard type has a white coat, with a black head and forequarters and a black stripe down the back. Since the pattern is variable, Castle used it to study what alterations could be produced by stringent selection. In one group he bred only from those with the most black; in another, from the most white. The results of this artificial selection were striking, for he was able to produce rats whose coat colours exceeded the normal range of variation. Some were nearly completely black; some were all snowy white with the 'hood' reduced to a mere smudge on the nose. This experiment was seized upon as a demonstration of the way selection, by working on small variations, can ultimately produce something that had not been seen before. 'Exactly the same thing can happen in Nature, only on a grander, looser scale and over slightly longer periods . . .' Julian Huxley proclaimed enthusiastically in *The Science of Life*, the popular book on evolution he co-authored with H.G. Wells and his son in 1929.

In the wild there had been no more than a handful of hit-or-miss Darwinian experiments. In 1894 the British biometrician W.R.F. Weldon gained an eternal footnote in evolutionary history by determining that in silted water natural selection favoured crabs with diminished 'frontal breadth'. A few years later, in the aftermath of a severe storm that hit the coast of Connecticut, a man named H.C. Bumpus found a number of English sparrows in distress on the ground. When he took them indoors and tried to nurse them back to health, some revived and some died. Bumpus was evidently a very curious man, for he took measurements with calipers and discovered that the dead sparrows generally had wings shorter or longer than the average length. Apparently, survival of the average, or 'stabilizing selection', was operating. But these proto-experiments, and others like them, had an anecdotal quality, and Fisher, along with Haldane and Wright, longed to infuse biology with the rigour of the physical sciences, to find within the rise and fall of alleles in woodlands or fields something approaching the lawlike regularities of Newtonian physics. Now that there was a mathematical theory of natural selection, natural selection ought to become measurable and testable. But who could test it? And on what?

Genetics in those days was an indoor pastime. The climate was controlled by central heating. At any time one could look inside the milk bottles and see how the *Drosophila* drama was unfolding, which phenotypes were the winners of life's competition. All the progeny were accounted for; none flew away, were blown away, or got eaten; none starved, unless on purpose. Studying an animal population in nature is a trickier proposition, for mating preferences, migration, crowding, hibernation, early frosts, insecticides, unidentified predators, larval viruses, competition from other species, and many other factors complicate the issue.

Before the 1950s there were few instances where the genetics of wild populations had been mapped. In the United States the Ukrainian-born Theodosius Dobzhansky (1900–1975), a leading architect of the Synthesis, forged a collaboration with Sewall Wright. A veteran of the Columbia 'fly room', Dobzhansky ventured to the high altitudes of the Sierra Nevada range to study sixteen strains of the fly *Drosophila pseudoobscura*. The strains differed in particular 'inversions' in large segments of chromosomes and were known by their initials: ST (Standard), Arrowhead (AR), PP (Pikes Peak), and so on. Dobzhansky recorded spectacular peaks and declines in the population of different strains over several years. He meticulously ruled out chance, mutation pressure or internally directed changes, and proclaimed that natural selection was almost certainly involved. The selective factor, or factors, remained unidentified, however. And, as the science historian William Provine quipped, '*Drosophila pseudoobscura* is an animal without an ecology.' It must have one, certainly, but no one knew anything about it.

While Wright and Dobzhansky were pursuing their collaboration in the US, another theoreticist/experimentalist bond was forming in Britain between Ronald Fisher and a fervent young Darwinist, Edmund Brisco Ford. In 1923 Fisher, thirty-three years old and already becoming famous, was a fellow of Caius College, Cambridge. Edmund Brisco Ford, known to his friends as 'Henry', was a pasty, bespectacled undergraduate at Wadham College, Oxford, where he was working under Julian Huxley on genetic physiology. One day Huxley happened to mention young Ford to the great statistician. 'Other people in his position might possibly have asked briefly about me,' Ford later reminisced, 'a few might have even invited me to go to see them. Fisher's reaction was different. The Fellow of Caius took a train to Oxford to call on an undergraduate.

Characteristically it did not occur to him to let me know that he was coming so I was out when he arrived and he settled down in my rooms to wait for me.'

The young E.B. Ford opened the door to his sitting-room to find a blue haze of pipe smoke, 'a thing which disgusts me', and a smallish, red-haired stranger with a 'rather fierce pointed, red beard' and a 'very white face' that reminded him of King George V. As Fisher got up and came toward him, Ford was struck by his eyes, 'hard and glittering like a snake's and seen through spectacles with lenses so thick they resembled transparent pebbles'. Fisher took his hand 'in a firm bony grip and, bending slightly forwards, gave me a momentary but most searching inspection. Then his face relaxed in a charming smile, the beginning of nearly forty years of friendship.'

It would be an intellectual union of uncommon fervour, and part of its bond derived from the fact that Darwinians in 1923 felt a little like the Christians in second-century Rome when they were being fed to the lions. They were a beleaguered minority. Ford would recall that he, Fisher, Huxley and Haldane, whom he abhorred, 'were at that time almost alone in this or any other country' in preaching the importance of selection *and* Mendelism.

In time E.B. Ford would invent a new Darwinian science, which he baptized Ecological Genetics. This he would define as the field of study that 'deals with the adjustments and adaptations of wild populations to their environments ... Indeed it supplies the means, and the only direct means, of investigating the actual process of evolution taking place at the present time.' This is not a modest claim, but Ford was never a modest man. With Fisher, he dreamed of a way of studying evolution not in the laboratory but *in nature* that was experimental, rigorous, and, above all, quantitative. No one had tried anything like it before, and perhaps it required an ego as void of self-doubt as E.B. Ford's to attempt it. His grandiose vision, merged with Fisher's, would lead directly to H.B.D. Kettlewell's legendary experiments.

Odd-looking, slope-shouldered, epicene in manner, with a voice so high-pitched he was sometimes taken for a woman on the telephone, E.B. Ford was one of those people who seem to have skipped the stage of childhood. His early years are shrouded in mist, like the legends of mythical heroes. We know that he was born in 1901, the same year his lifelong idol, Queen Victoria, died – and all his life he would retain something distinctly Victorian. The place of his birth was the Lake District, specifically, Papcastle, Cockermouth, Cumbria. He liked to hint at vaguely aristocratic origins, and listed among his ancestors Catherine Howard, one of the wives of Henry the Eighth, but it was hard to be quite sure, for Ford was an incurable snob and dropper of names. Although he spoke fondly of his father, who was a vicar, a perfectly respectable middle-class calling, he always avoided mentioning his profession.

He was an only child, and quite likely a very lonely one, and he took up the pursuits of a bright, lonely child. By collecting Roman coins that turned up around the Manor House in Papcastle where he lived, he began dabbling in amateur archeology. He read the classics. With his father, he collected butterflies, becoming a convinced Darwinian and an amateur scientist while still in his teens. He and his father did fieldwork on a butterfly, *Melitaea (Euphydryas) aurinia*, the marsh fritillary, which gave him his first clue about the rate of evolutionary change. It had become very rare when they first looked for it in a certain location in Cumbria in 1917, but suddenly, around 1922, they found a 'dancing haze' of *Melitaea* butterflies of dizzying variety. 'Hardly any two were alike . . .' he would recall. 'Those that differed most were completely abnormal and couldn't expand their wings properly.' A fundamental article of his faith was that evolution was accelerated under conditions where many variations were being produced, and *Melitaea* became a sacred story he invoked ritually to explain his personal mission, never failing to recall a pivotal conversation with Leonard Darwin. 'When he told me later that his father [Charles Darwin] thought it would take about fifty years to detect and study evolution in an annual species I, remembering *M. aurinia*, felt sure he was wrong.'[25] Ford thought evolution could happen much faster than the old Darwinians believed, and eventually he would prove it.

Within his family Ford was always 'Edmund', and no one knows when or how he metamorphosed into 'Henry', or whether he named himself after the American automobile tycoon or someone else entirely. By the time he turned up at Oxford, whatever his motives he was called Henry Ford. He knew already that his vocation was 'the study of evolution by means of observation and experiment, an interest that derived from our work on *Melitaea*'; but there was no such field of study in early Twenties Oxford. There was precious little genetics either. As the lone zoology undergraduate researching genetics, he was given an amphipod crustacean, a sort of brine shrimp called *Gammarus chevreuxi*, to study. The idea was to breed it and learn how the genes segregated in a Mendelian manner, but Ford became riveted. He went on studying the animal as he began to suspect that the genes for eye colour might control the rate of development. He teamed up with Julian Huxley, then a lecturer, to research *G. chevreuxi*, but in the midst of this very complex series of experiments Huxley left Oxford to take up a post at King's College, London, leaving the undergraduate to carry on by himself. Later he would recall how frightening this was.

One thing the humble shrimp had to offer was eyes ranging in colour from black to chocolate, to reddish brown, to red. The colour, caused by the rate of deposition of melanin, could be modified by temperature but was also subject to a complicated genetic control. This pointed to the very important idea of 'rate genes'. Ford discovered some rate genes that slowed development, and also demonstrated that the growth of the body as a whole affected the eye colour. It was a landmark study of the interplay of genetic and environmental factors in controlling the expression of a particular gene, and when it came time to write it up for publication, Huxley was magnanimous enough to allow the undergraduate to be the first author on the paper. E.B. Ford had earned his spurs.

Ford was a lifelong bachelor with a deep-seated antipathy to the female sex, with a few prominent exceptions; he was precise, methodical, and always impeccably dressed. Fisher seemed to be his polar opposite, a rumpled, distracted married man with a sprawling family of six girls and two boys. However, at his core Fisher remained as solitary as Ford. During their marriage, his long-suffering wife Eileen, in between bearing and nursing their numerous eugenically sired progeny, was obliged to cater to her husband as to an Asian prince. She played 'medicine ball' with him before breakfast, read to him, polished his boots, kept him company when he suffered from his frequent headaches and insomnia, helped to build cages for his experimental mice. His daughter would recall that 'with the egotism of a child, Fisher assumed that he came first with his wife in all the circumstances of life' and he would accuse her of not loving him if she took a much-needed nap under doctor's orders. In his anthropometrical zeal, he brought home calipers to measure the length and breadth of his children's skulls, and the older children were brought to his laboratory to take a battery of tests. In 1943, his son George died in a plane crash, and Ronald searched fruitlessly for a young woman he might have impregnated beforehand and to whom he might have passed on his genes. Grief-stricken, he abandoned normal domesticity altogether and moved, with his mice and dogs, into bachelor's quarters at Gonville and Caius College, Cambridge. When the college offended him by appointing a woman bursar, he moved in to the top floor of the genetics department in Storey's Way. Having successfully passed on his genes, he never returned to family life.

The botanist David A. Jones[26] was one of Fisher's last students before his mandatory retirement in 1958. He recalled: 'He was absolutely marvellous as a teacher. And he could multiply ten- or eleven-digit numbers in his head if he had to. His only problem was that he would assume that everything he'd said in previous lectures had stuck. There was a marked decline in attendance after the second lecture.

'The first day he announced that the five o'clock lecture would begin and end five minutes early because there was a five-minute walk to The Bun Shop, a pub that opened at six o'clock. At opening time Sir Ronald Fisher would be there in the pub, and he'd sit with a pint for three-quarters of an hour, talking to students. He was like Aristotle, like the peripatetic philosophers. Then he would bicycle to Caius College for his evening meal at seven. He was terribly dangerous on his bicycle, notorious for running people down.'

Through E.B. Ford, Fisher became interested in field data related to the genetical theory that was forming in his head. In 1926 the two men published a study of thirty-five species of British moths, determining that more abundant species had more genetic variability than the rarer ones. This confirmed a prediction Fisher had made in 1922, and he became smitten with the potential of fieldwork. Using statistical tools developed by Fisher, the two men carried out field studies on primroses and butterflies. Fisher was so myopic he could barely see a butterfly, much less catch one, and was physically inept, but his mind seemed nearly godlike to Ford. In those days it was still an uphill battle to show that natural selection occurred at all and that it could direct evolutionary change. Hence Ford was grateful for Fisher's mathematical proof that very small selective advantages could fuel evolution – an insight he never ceased citing in reverential tones, even though he believed selective pressures were actually greater than Fisher theorized. (In this Ford agreed with Haldane, but such was his personal loathing for him that he rarely cited him.)

A misogynist of ancient Greek proportions even by the standards of his day, Ford nonetheless worshipped several women. The first, apart from his lifelong idol Queen Victoria, was Evelyn Clarke, to whom he usually referred as 'my friend Mrs Clarke, FSA [Fellow of the Society of Antiquaries]'. An elegant, straight-spined society lady, married to a scion of the Clarke shoe family, she seems to have been a much older cousin of Ford's, though he sometimes referred to her as his 'aunt', and it was in her home in Street, Somerset that he spent every birthday (23 April) and Christmas and other family holidays. He always reverently signed her guest book, and as he sat stroking her cat in his lap he might be heard to murmur: 'Pussy, my love, suppose we all bit and scratched like you: what would Queen Victoria say?' Mrs Clarke and 'Edmund', as she always called him, had a lifelong bond, cemented by a shared passion for amateur archeology. Together they unearthed important Iron Age sculptures in a Somerset fogou, or underground chamber, publishing a scholarly paper, 'The Fogou of Lower Boscastle, Cornwall', which Ford was fond of citing, partly because he loved uttering the word 'fogou'. Whenever someone was ignorant or gauche enough to need to ask: 'And Dr Ford, what do you do?' he always answered: 'Archeology.'

Another 'female woman' who transcended her sex in Ford's

eyes was Miriam Rothschild, who at the time she met him went by her married name, Mrs George Lane. She was a Rothschild by birth, a niece of the collector Baron Walter Rothschild, a circumstance that by itself would have whetted Ford's social-climbing palate; but she was also a highly intelligent self-taught entomologist, an expert on fleas, who, despite having no degree, was soon to become a *de facto* member of the Oxford School. Eventually she would earn eight honorary doctorates, including those from Cambridge and Oxford, and become a Fellow of the Royal Society, as well as authoring several well-respected books and being a force for conservation, homosexual rights, and several other causes. 'She has very much the mind of a man!' Henry would say admiringly. After corresponding with him about butterflies for several years, 'Mrs Lane' came to visit Dr Ford for the first time in 1956, knocking on the door of his office punctually at 11 a.m., as he had suggested.

After a moment's silence there was rather a plaintive long drawn out cry: 'come in!' I opened the door and found an empty room. I looked round nervously – not a soul to be seen, but an almost frightening neatness pervaded everything. Each single object, from paper knife to *Medical Genetics*, was in its right place . . . the sight of this distilled essence of neatness and order took my breath away. I stood there, probably with my mouth open, trying to reconcile this vacant room with that ghostly cry – had I dreamed it – when suddenly Professor Ford appeared from underneath his desk like a graceful fakir emerging from a grave. Apparently he had been sitting cross-legged on the floor in the well of his writing table, lost in thought, but he held out his hand to me in a most affable manner. His explanation for this rather startling welcome was: 'My dear Mrs Lane – I didn't know it was you.'[27]

Although immensely conceited and domineering, Henry could be exceedingly thoughtful at times, Dame Miriam recalls. Her secretary had committed suicide, leaving her two orphaned children for Rothschild to raise along with her own four. As she was feeling overwhelmed about caring for six children under the age of ten, Ford came to visit and remarked: 'Mrs Lane, you must have some help. You need a governess!' 'I said, "Professor Ford, they don't exist any more." He said, "It depends." Within twenty-four hours he had a governess on my doorstep, a Mrs Brown.'[28]

No one could be more solicitous and thoughtful to favoured friends or protégés; no one more casually caustic to telephone operators, waiters, out-of-favour students, or others who fell into the category of the not-approved. 'Hopeless girl!' he would pronounce frequently of ungifted female students or assistants. 'If you do not look at its face, I think you will be able to tolerate it quite all right,' he once wrote of a young man applying for a job. 'The thing is rather a streak of misery . . . It did smile once, in a rather sickly way.' Wives of graduate students were a particular inconvenience. 'He is so foolish as to bring a wife with him, having just married . . .' he wrote, characteristically, of a new graduate student. 'I am having them at Apsley Road (the wretched female-wife precludes having them at All Souls) for the first two nights.' He treated them with elaborate courtesy mingled with subtle intimidation.

Perhaps it was inevitable for one who saw men and women as fundamentally discontinuous forms that Ford's speciality should be polymorphism. He would begin each and every lecture by reciting the definition in his high-pitched, languid voice, as if it were the definition of God in the catechism. A polymorphism, Ford would intone, is:

the occurrence together, in the same locality, at the same time, of two or more discontinuous forms of a species in such proportions as the rarest of them cannot be maintained by recurrent mutation.

In other words, there must be more of each type than could be produced by the odd mutation alone; their existence must be maintained by natural selection. He would go on to explain that the definition excluded continuous variation, geographic races, seasonal variations, and 'sports', occasional rare gross genetic aberrations.

Genetic variations come in two forms, continuous and discontinuous. In discontinuous variation, or polymorphism, the difference between two characters is sharply delineated (a pink shell versus a brown, two spots on a wing versus four spots, a black peppered moth versus a pale one). Because polymorphisms conform to the one-gene model of genetics and can be analysed in a simple Mendelian fashion, they are tailor-made for testing the theory of natural selection.

Some polymorphisms are more apparent than others. Eye and hair colour in humans are clearly visible whereas blood-group, enzyme and taste-blindness polymorphisms require special tests to be detected. At one time Ford sampled Oxford undergraduates for the ability to discern the scent of freesias, a polymorphism possessed by a minority of human beings, and he collaborated with doctors to collect blood and saliva samples in an Oxford maternity hospital. It is hard to picture the famously uncuddly E.B. Ford swabbing out the mouths of infants with squares of cotton wool, and sitting at the bedside of new mothers interviewing them about previous pregnancies and miscarriages, but this is what he did in the early 1940s. He was testing for the rhesus factor in blood, a human polymorphism of medical importance.

Polymorphisms had been problematic for the early Darwinians, because it seemed that 'survival of the fittest' should lead to *the* fittest form, not two or more. Darwin believed polymorphisms had nothing to do with natural selection, but Ford and Fisher thought otherwise. Alleles, or allelomorphs, are different forms of a particular gene – one that codes for black wing colour, say, and the other for pale wings – and Fisher's mathematics showed that two or more could persist in a population if they were maintained by a *balance* of selective forces.

From the operation of natural selection, Ford and Fisher perceived, either a 'transient polymorphism' or a 'balanced polymorphism' could result. If selection favours one of two phenotypes, the polymorphism will be shortlived because one allele will be eliminated eventually and the alternative fixed in the population. However, under certain conditions the polymorphism will be balanced and can continue virtually forever. It is also difficult to eliminate a harmful recessive allele from a population, since, being unexpressed, it can persist almost indefinitely, buffered against selection. Natural selection could have three different outcomes for a population. Normalizing or stabilizing selection favours the average individual in the population and eliminates the extremes. (This was presumably the case with Bumpus's sparrows.) Progressive, or directional, selection favours one extreme – the darkest coat, let's say – and so over time the population's mean will shift. Disruptive selection favours both extremes – say, very long beaks and very short beaks – and eliminates average individuals, so that over time the population bifurcates into two separate distributions.

Although E.B. Ford did not invent the concept of polymorphism, he defined it for all time and claimed it as his own. One obvious, tantalizing polymorphism, he was aware, existed in the peppered moths of England.

The genetics of *Biston betularia* are reassuringly Mendelian. The *carbonaria* gene (C) is dominant; the typical gene (c) recessive. A cross between a melanic (CC) and a typical (cc) can produce C C (a *carbonaria* homozygote), Cc (a heterozygote, with one of each allele), or cc (a typical homozygote). The genes actually code for the presence or absence of melanin in the wing scales. Because C is dominant, the Cc heterozygotes will be melanic. A segregating brood will have, on average, the ratio 1 C C, 2 Cc and 1 cc, and in appearance these will be three melanics and one typical. (In addition to the *carbonaria* and *typica* forms of the peppered moth, there are three different alleles for intermediate forms known as *insularia*. Because of the complexity and uncertainty surrounding them, we will, with apologies to *insularia*, give them scant attention in this story.)

When Ford turned his attention to this famous moth, it was with the intention of correcting the errors of people whose views he deplored.

If you were keen on evolution in the 1930s, you might have owned a copy of *The Science of Life* by Julian Huxley and H.G. Wells, an encyclopedic collaboration for which Huxley quit his post at King's College and engaged two full-time secretaries for several years. While he knew less biology than his co-author, Wells had the good sense to censor such Huxley outbursts as the following: 'It comes to this, that the evils of slum-life are largely due to the slums, but to a definite extent caused by the type of people who inevitably gravitate and will make a slum for themselves if not prevented.' Wells pleaded in a marginal note: 'It could be. I pray you not to do [write] it.' Published in 1929, the book is a fascinating period piece, containing just about everything then known about evolution, reduced to bite-size pieces. Looking up 'industrial melanism', you would have found, after a brief account of the rise of industrial melanism, the confident statement: 'No help in escaping the notice of enemies seems to be conferred upon these moths by their dark colour; it is not a protective variation.' The entry continues:

Now all green things in industrial districts are coated with a grime that is rich in poisonous metallic salts. It occurred to [the biologist John Heslop] Harrison, struck by the coincidence between the distribution of black moths and of industrial smoke, that this might be the cause of the change. Accordingly he made the caterpillars of various forms eat tiny quantities of heavy metals, especially lead and manganese, with their food . . .[29]

A full ten years earlier, Professor John W. Heslop Harrison of Newcastle University had reviewed previous work on melanism and vehemently rejected J. W. Tutt's hypothesis, articulated in 1896, that the rise of industrial melanism depended on selective bird predation. Instead, he said, melanism was caused by some effect of chemical pollutants on the moth and on its genes. He reasoned that industrial fallout coated leaves, and when caterpillars ate them the metallic salts in the pollutants turned the insects dark. The pollutants affected not only their bodies but also the 'germ plasm' (the genetic material), so that the melanic trait was passed on to offspring.

Heslop Harrison set out to produce heritable melanic forms experimentally, and he fed larvae of different species (not including *Biston betularia*) on foliage coated with some of the foulest industrial byproducts around, lead nitrate and manganese sulphate. His results were announced in several influential papers, starting in 1926. Here is how Huxley and Wells summed them up:

His suspicion was justified; in the metal-fed cultures a few mutants with black wings appeared. Moreover, the colour, once it had been produced, bred true without further metal-feeding. The chemical agencies had induced permanent changes in the germ-plasm . . .[30]

The fact that this popular book on evolution, by Huxley and Wells no less, touted Harrison's theory and failed to cite Tutt is a measure of how things stood in 1929. Although Harrison was criticized for obtaining results in species that did not show melanism in the wild, and although he had not worked on the peppered moth, his research dominated the field of industrial melanism entirely. Other scientists failed to replicate his experimental results, but Harrison stubbornly insisted that his critics' techniques were flawed, as he would continue to insist until his dying day, long after rumours of fraud surrounded his name.

It happened that Ford disliked Heslop Harrison intensely. This was not entirely surprising, for Ford loathed many people, and Harrison inspired dislike in quite a few. Having risen from a modest background as an ironworker's son in a small village in the north of England, Harrison had passed through one of the 'redbrick' universities rather than Oxford or Cambridge, and he carried a massive chip on his shoulder. He was prickly, abrasive, domineering, intolerant of criticism, and inclined to feuds.[31] E.B. Ford, educated at Oxford, mannered as an eighteenth-century courtier, bristling with upper-class affectations, was also prickly, abrasive, intolerant of criticism and inclined to feuds. The two men were more similar than either would have been willing to admit – like doppelgängers seen in a funhouse mirror – and they had in common a habit of referring to important people as 'my friend X'. Ford, who dropped the names of many important personages among his 'intimate friends', was once heard to refer to 'my friend the Pope'.[32]

It was a feud made in heaven. Perhaps partly to rankle

Heslop Harrison, Ford put natural selection back on the table in 1937, in a theory of industrial melanism he would elaborate over the next few years. The melanic form of the peppered moth, he proposed, was hardier than the typical. As a byproduct of his famous theory of dominance, Ronald Fisher had predicted that heterozygotes (those with two unlike alleles) would be generally fitter than homozygotes, and for Ford just about every Fisherian utterance was gospel. Among peppered moths, the heterozygotes are melanics, indistinguishable from homozygous melanics. (The melanic allele is dominant, concealing the recessive typical allele.) Their superior hardiness would give melanic moths the edge under the harsh conditions of an industrial environment, Ford theorized. In the rural countryside, however, the black moths would be more conspicuous and would be plucked off the tree trunks by birds, nullifying their advantage in viability. He agreed with Tutt that crypsis and avian predation would play a role.

At the time E.B. Ford was reapplying Darwinian theory to the peppered moth, the great marriage between evolution and Mendelian genetics had been barely consummated. As a Darwinist with a degree in genetics, Ford was himself a child of the Synthesis – one of the first Synthesis babies, in fact – and he was as messianic as a prophet who has had a brief and tantalizing glimpse of the promised land. He knew what needed to be done. The fact was that a malaise still haunted Darwinism even in the 1930s. While the theoretical framework of the new science was in place, or would be as soon as all the founding fathers finished writing their manifestos, a rigorous experimental science was still lacking. There was, in short, much theoretical talk about very sparse evidence. It would be necessary to go out and prove T.H. Huxley wrong about the futility of proving evolutionary theory; necessary, moreover, to prove Darwin wrong, for Ford was sure that some evolutionary changes were observable within the span of a man's lifetime. It was just a matter of knowing where to look.

In 1937, just as a war in Europe was beginning to loom, Henry Ford saw that industrial melanism might serve to demonstrate the effect of natural selection in *nature*, as opposed to the artificial selection of the laboratory. But he would need a gregarious, larger-than-life moth lover, Henry Bernard Davis Kettlewell, to do the hard work.

4

The private lives of insects

T he Black Wood of Rannoch stands in the middle of a remote part of the central Scottish Highlands, fringed by peat bogs and hills that grow larger and more severe to the west in the approaches to Glen Coe. The area attracts few casual visitors – its sights are for hardy outdoor enthusiasts only. It is interesting to wonder, if E.B. Ford had not bumped into Kettlewell in the Black Wood of Rannoch, whether the biology texts of the next half-century would be stripped of their persuasive power. Perhaps someone else would have come forward to perform the peppered moth experiment. Or perhaps biology students would be learning of another legendary experiment, on an entirely different species, as 'Darwin's missing evidence'. We cannot know, but if we were metaphysically inclined we could make a case for an occult synchronicity, a form of dream logic, in the fact of their meeting in the same year – 1937 – that Ford first hazarded a theory of industrial melanism.

The primeval Black Wood of Rannoch was a collector's El Dorado, an unusual tract of Caledonian pine forest, unique in the British Isles, where rare species abounded. On this particular day Ford and a lepidopterist colleague were searching for a particular butterfly; we no longer know which. To find the rare species you covet, you must know exactly when and where to look, and collectors are famously stingy with this type of information. As it happened, Kettlewell didn't mind giving Ford and his lifelong collector friend 'Bunny' Dowdeswell directions to the correct locality, for which Ford pronounced himself 'immensely grateful' in a subsequent thank-you letter. He was an inveterate writer of unctuous thank-you notes.

What Ford would have seen before him was a tall (six foot four), sun-tanned, boisterous thirtyish man in shorts, with sunburnt knees and heaps of bonhomie. As Kettlewell spoke with his chin in the air in his loud, jocular voice, plumes of smoke from the cheroot clamped between his teeth would have swirled into the crisp, pine-scented air. Kettlewell would have been met with a severe-looking bespectacled man on the verge of middle age, wearing neatly pressed khakis, and perhaps a tie and a trilby hat – E.B. Ford's fieldwork garb. His strange moon face, cadaverous grin and mannered speech gave him the air of a traveller from an earlier century, and he may have moved awkwardly in his body as if unfamiliar with the mechanism. It would not have surprised Kettlewell to hear that Ford was a don at Oxford, a pioneer in genetics; while Ford would not have been astonished to learn that Kettlewell was a medical doctor, a thirty-one-year-old general practitioner in Cranleigh, Surrey, south of London, who had left his wife and tiny daughter at home to indulge his passion for butterflies and moths.

There must have been some uncanny sympathy between the practice of medicine and entomology, because many of the contributors to journals like *The Entomologist's Monthly Magazine* were doctors. They sent in considered and meticulous observations, like this one from Chipstead, Surrey. 'On visiting a small, swampy pond I found 5 females and 1 male of *S. flaveolum*. One pair was taken flying "in tandum" [*sic*] and subsequently released, later observed in the act of oviposition.' Today, even in England, such patient observation of nature has largely been eradicated by satellite dishes and the internet, but this was the milieu of Bernard Kettlewell.

From the chance meeting in the Black Wood blossomed a typical lepidopterist pen-pal relationship, in which Ford and Kettlewell began to trade breeding tips and to send each other 'material' – eggs, larvae, butterflies – through the post. 'I am interested to hear of your temperature experiments on Heliothis peltigera. I wonder if you find any indication of a critical period at the start of pupation, as Suffert records,' Kettlewell was writing Ford soon after their meeting.[33] The letters started formally, but before long they were 'Henry' and 'Bernard' to each other. Bernard Kettlewell was the prototype of the amateur bug-hunter. Like most, or perhaps all, lepidopterists, he had acquired his hobby in childhood. An only child, like Ford, he was born in Howden, East Yorkshire, and raised in Birmingham. Bernard's father, who was tiny as Bernard would be huge, was a member of the Birmingham corn exchange and 'always acted as if Bernard was a sort of mistake', according to Bernard's son, David. Bernard's domineering mother demanded complete compliance and made her son her lifetime personal project. Every time he came home he would find her waiting for him on the landing at the top of the stairs, demanding a complete accounting. 'He was terribly afraid of his mother but also desperately attached,' according to David.

At the ancient Charterhouse School, in Surrey, Bernard was remembered as an undistinguished student with a passionate, unruly nature, prone to practical jokes. Charterhouse had a naturalist and scientific bent (one of the masters who had influenced Bernard was the remarkable George Mallory, who would perish near the summit of Everest in 1924), and it had moth traps set up in the woods. Partly to escape the Spartan discipline of school life, Bernard took to visiting the traps often, and he began to know his moths. Before long, he was smitten, a common formative experience of moth men. Like a pining suitor, he would climb out of his dormitory window at night and bicycle through the dark to the nearby town of Guildford to catch the moths hovering around the streetlamps. When he was caught he was beaten, but nothing could deter him from his hobby.[34]

Non-entomologists picture 'butterfly hunting' as the most serene and uncompetitive of sports, somewhere between crocheting and backgammon, but nothing could be less true. The drive to be the first to 'take' a particular variety (the vaguely sexual overtones are no accident) or to find its breeding ground rivals the cut-throat avidity of prospectors in California in 1849. The old entomology journals are crammed with bitter rows, accusations, broken friendships, wounded egos and desperate undertakings, all punctuated by salvos of taxonomic Latin.

As a teenager hunting for caterpillars, Bernard would launch himself into fields and bogs with reckless ardour, pushing himself to find twice as many as his companions, and would quarrel violently over disputed discoveries. His gift for natural history shone early on, according to his boyhood friend Ronnie Demuth. After beating the bushes of the New Forest for larvae for days, the schoolboys at last caught a great rarity, an alder moth – *Apatele alni* – larva in its penultimate instar, when

71

it resembles a bird dropping. 'Why does it mimic a bird dropping?' Bernard wondered aloud. 'Because it sits exposed on the top surface of leaves [where its resemblance to bird-doo would fool predators]'. This was a clue to the caterpillar's locality. Sure enough, when the boys bent the slender saplings bordering the trail, the upper leaves were acrawl with prize *alni* caterpillars. That night, Bernard and Ronnie ran into a well-known collector, obviously vexed to find his site trampled by schoolboys.

'What have you boys been collecting?' he asked.

'We have got a lot of *alni* larvae.'

'You must be mistaken,' the eminent collector replied condescendingly. 'I have collected in the Forest for twenty years and have never seen one yet.' The next morning, Bernard and Ronnie visited him in his campsite during his breakfast, whipped the lid off a biscuit tin, and exclaimed: 'We have come to show you the *alni* larvae,' all of which 'spoilt his breakfast'.[35]

With moths and butterflies always in his thoughts, Bernard longed to make entomology his career, but his mother was dead-set against it. 'Bernard, you cannot make a living this way!' she would tell him. Surrendering to his mother's implacable agenda, he read medicine at Cambridge and, because his heart wasn't in it any more than Charles Darwin's had been, suffered a 'mini nervous breakdown', in his son's words, at the age of twenty-one. As a 'medico' he would always have the mildly haunted air of a man who has made a marriage of convenience but never stops dreaming of the woman he really loves. With fellow students in the entomological section of the Cambridge University Natural History Society, he frequently obtained permission from his tutor to spend the night bug-hunting in the Fens, and his fast car allowed him and a few friends to range even farther afield in search of rarities. It did not take him long to rack up lepidopteral triumphs, which were recorded in the entomology journals with a *gravitas* appropriate to military exploits, even while he went about the business of earning an MA in zoology and a medical degree. He also collected pretty girls during this time: Bernard always had an eye for women, and any party was an instant success the moment he walked in the room.

During the time he was headquartered in London in the 1930s, gaining his medical qualifications at St Bartholomew's Hospital, he spent his summer months driving at breakneck speeds through the flat green expanse of East Anglia in his racy Alvis convertible, snaring rarities of the Fens. 'At high speed the brakes would go on with a squeal,' recalled his friend Demuth, who went on to share a staircase with Bernard at Caius College, Cambridge. "Plover nest!" The car would violently reverse and Bernard would get out and stride a surprisingly long way across an adjoining field, unerringly to the nest, and return with the eggs. "They will do for *my* breakfast".'[36] In 1936 Bernard married a stunning society girl, Hazel Wiltshire, the daughter of the town clerk of Birmingham. A slim and elegant brunette, charming, spirited, high-strung, and something of a clothes-horse, she had grown up surrounded by wealth and servants. She had the skills of a great hostess, and when she married Bernard she believed she was assuming the role of a prosperous medical doctor's supportive wife. Like Bernard, Hazel loved entertaining, and excelled at it, and their home was always overflowing with house-guests. In 1937 their first child was born, a daughter, Margaret Dawn, who would always be known as Dawn and who was destined to cause her parents a world of heartbreak; in 1944 a son, David, followed. Marriage and fatherhood did nothing to dampen Bernard's ardour for the woods, fields, streams, campsites, and the vertiginous variety of winged creatures, and Hazel graciously put up with it all.

While still an undergraduate at Cambridge Bernard met one of the pivotal people in his life: the entomologist E.A. Cockayne, who was an *éminence grise* in natural history circles. A celebrated medical doctor (with an OBE after his name), he was a severe and intimidating bachelor who lived alone in a great house, terrorizing his servants with his difficult moods. Although he was an amateur, scientists would be citing his papers reverently for decades. Recognizing Bernard Kettlewell's abilities, Cockayne began to groom him as his scientific heir. When they found a rare butterfly or moth, they would breed it to another variety, often producing varieties that had never existed before. They recorded the genetics, tracking three or four generations in their 'brood books'. They wanted to know if such and such a trait was genetic, and if so, was it determined by a single Mendelian factor or several? Was the gene dominant or recessive? They did temperature-control experiments to determine whether raising the pupae at 30 degrees centigrade would affect the wing markings of the moth that would emerge. They founded artificial colonies of certain butterflies. They were *almost* scientists, or rather, they were not unlike the gentlemen-scientists of the eighteenth and nineteenth centuries.

Cockayne agreed to merge his considerable collection with Kettlewell's and, united with the collection of Lord Walter Rothschild, this became the Rothschild-Cockayne-Kettlewell Collection of British Lepidoptera (later renamed the National Collection), donated to the British Museum of Natural History. Once they became partners, the two men found themselves locked in an intense, turbulent, bickering bond that both would liken to a marriage. Cockayne had a habit of browbeating the younger man, scolding him for killing a butterfly before its wings were properly dried or for making careless errors in his brood books. ('Books properly kept give pairings and results in full . . .') Even Bernard's published work failed to please. ('I've read *Nature* and don't like your letter.')[37] Perhaps Bernard's childhood had made him peculiarly susceptible to judgemental parental figures whose approval he was doomed always to try, and fail, to gain. Certainly, there is much in the relationship with Cockayne that prefigured the one he was destined to have with the great E.B. Ford, and, had Bernard consulted a fortune-teller at a fair, he might have been warned off. At first, however, Ford and Kettlewell were as delighted with each other as any two monomaniacal collectors could be.

By 1939, Henry Ford had attained the title of Lecturer in Genetics. He was famous at Oxford for his ability to lecture fluently without notes – holding forth in complete, grammatically flawless sentences as if he were reading a script off the back of his eyelids. 'He was superb,' remembers Michael Majerus, who heard him twice. His rare ability to explain difficult concepts clearly made his lectures very popular. It was important to arrive on time, however. 'Due to the shape of the lecture theatre, anyone who came in late was seen by the lecturer first,' recalls David Jones. 'Ford had a marvellous command of the language. Just at the point where the heads started to turn toward the latecomer, he would insert into his lecture an extremely long word. He would pause in the middle of the word as the latecomer moved toward his seat, then when the person sat down, finish the word as if nothing had happened. It was very effective.'

Ford spent the war years doing seminal work with Fisher on polymorphic butterflies and primroses, elaborating his theory of the medical significance of the polymorphic human blood groups, and generally establishing his preeminence in the field. In 1946 he was made a Fellow of the Royal Society. (He would go on to win the Society's prestigious Darwin medal in 1956.) The next step would be the baronial position of professor, and he must have been in a fever to become 'Professor Ford' instead of merely 'Dr Ford'. At Oxford the title of professor applied only to heads of department, which, in his case, narrowed the possibilities down to two: the Linacre professorship of Zoology or the Hope professorship of Entomology. The Linacre professorship was already occupied, by Alister Hardy, his boss, leaving the Hope professorship, just vacated in 1947. Ford's papers reveal that he campaigned assiduously for the Hope professorship, but the prize went to a rival, who thereby earned his eternal dislike. Ford let it be known that it was he who had 'decided against' the Hope professorship, disliking the curating duties, but that he had managed to procure something better: his own subdepartment within the department of zoology. This little fiefdom would be called the Oxford School of Ecological Genetics, and all his life Ford would talk about it as if it were a real department, although 'it existed only in the mind of God – or E.B. Ford', as a former student put it. Formally, it was a part of the department of zoology, and depended on the generosity of Professor Alister Hardy for its allocated space within the University Museum on Parks Road.

A fine example of High Victorian Gothic architecture, the museum had a soaring vaulted glass roof supported by ornate wrought ironwork; in the Victorian spirit, its columns had capitals and bases carved with naturalistically rendered groups of flora and fauna. In 1860, one of its lecture theatres had hosted the Huxley–Wilberforce debate on evolution. On the way to their second-floor offices Ford's bright young men (and the occasional bright young woman) would stroll past dinosaur bones, molluscs and crustaceans, rows of dusky beetles and powdery pinned butterflies, perhaps pausing before the forlorn spiky claw and long-beaked skull that are the remains of one of the last dodos on Earth. Up the wide staircase, their laboratories exuded earthy and acrid animal smells, and a visitor might hear the faint squeaks, crics, rustlings and warblings of mice, caterpillars and birds. Nearby was the Hope Department (which Ford persisted in calling the 'Hopeless Department'), with its high vaulted ceilings, banks of polished mahogany cabinets, out of which slid drawers of pinned Lepidoptera, and its fine library, ever useful to the lepidopterists in Henry's department.

Ford immediately recruited a handful of promising graduate students. Most of 'Ford's boys' came there with firsts (highest honours), usually from Oxford or Cambridge, and the brightest star in this firmament was the young lepidopterist Philip Sheppard, a returned prisoner of war. Shot down over the North Sea on his twenty-first birthday, Sheppard had spent the remainder of the war at Stalag Luft III, a German POW camp for downed British and American airmen 100 miles southeast of Berlin, and had participated as an 'earth bearer' in the heroic 1944 escape attempt later Hollywoodized as *The Great Escape*. In order to dispose of the two hundred tons of sand dislodged by the digging of escape tunnels, Sheppard and fellow prisoners would fill long, thin bags with sand, slip them inside their trousers, and then slowly let the sand trickle out while walking about the compound. The escape plan was elaborate and ingenious, involving forged identity papers and compasses constructed from fragments of gramophone records and magnetized razor blades, but it was foiled by botched signals, snowy weather, and the alertness of German sentries. On Himmler's orders, the fifty escapees recaptured (only three eventually made it to freedom) were handed over to the Gestapo and shot, an aftermath about which Sheppard could be coaxed to speak only after several beers. While in the camp he had been permitted to take a correspondence course in lepidoptery, and his resourcefulness would serve him well in Ford's department, where one had to be independent and self-sufficient to survive and where approximately one-third of the candidates washed out before finishing. 'When I came there as a graduate student, they said, in effect, "Here's your room. Get on with it,"' recalls Bryan Clarke.[38] This aloofness was Oxford policy, not just Ford's, and was understood to embrace the possibility of failure. During the oral examination of one doctoral candidate who was having difficulty with his thesis, Ford would recall wryly, 'we had a discussion, and while we were having the discussion, he was committing suicide in the other room.'[39]

In general Henry behaved as if he were auditioning for the Great Book of Eccentric Oxford Dons. People who knew him find the temptation to mimic his mannerisms irresistible. The eyes close halfway, as if world-weary. A sort of asthmatic throat-clearing precedes the first word. Then the unmistakable voice: high-pitched, mannered, smooth as oil, every syllable articulated, each vowel elongated. It had the effect of a High-Church sermon, with a tinge of something more louche. When I finally heard Ford's voice on tape, it was as if I had already met him, so thoroughly had his speech patterns been imprinted on his hearers. Many E.B. Ford stories persist in the Oxford memory stream, becoming highly polished bijoux in the retelling. There is a consensus on which is the most famous.

It was Ford's custom to give a zoology lecture to Oxford undergraduates twice a week. One day all the male students were absent for some reason, and only the women turned up. Ford trotted in to the lecture theatre, scanned the audience and drawled: 'Since no one is here today, the lecture is cancelled.' In another version, the students had arranged for all the male students to stay away, except for a lone man who did not get the message. Ford ran his myopic gaze over the assembled group, and said to the slide projectionist: 'I see only one person is here today. Therefore the lecture is cancelled.' There are other variants, and the legend appears in several books about Oxford.

Laurence Cook, who was a graduate student in the Oxford School of Ecological Genetics in the early 1960s, first heard the story from someone who had been there in the 1940s. He heard it again from a student from the late 1970s. 'He said that the undergraduates used to arrange things so that there was an all-female class, and Henry would come in and say, "Since there is nobody here . . ." They would set him up to do it so they could all giggle. That means he's been playing the same joke for thirty-five years. If he did it once it's funny, but if it happened over and over it's sort of a sad story.'

In daily life Henry's mode of expression was idiosyncratic or archaic. He travelled by 'motorcar' or 'flying machine', and any sort of device was, for him, an 'engine'. He pronounced Sellotape (Scotch tape) as if it were an ancient Greek muse, *sell-o'-ta-pee'*. Sometimes his language was so stilted he seemed to be translating from another tongue; some odd turns of phrase ('It hopes me that') suggested that he wished English had a dative form, like Latin and Greek. He frequently pronounced himself 'grobulated' or 'covered with shame'. Those who had witnessed his 'funny walks' could still re-enact them forty years later. If he wanted to signal the world that he was deep in thought and could not be disturbed, he would walk down the corridor listing to one side as if afflicted with a rare neurological disorder. Sometimes he could be seen doing what some called his 'wounded partridge' walk, hunched forward stiffly, with one arm bent behind his back like a broken wing. More often than not, he was muttering to himself.

Once he was seen to approach an elderly Hope Department technician and drawl: 'Taylor, I'm talking to myself too much.

You shouldn't talk to yourself, should you? If you hear me talking to myself, you will tell me, won't you?' Several days later he was observed babbling to himself while climbing the staircase to his office, Taylor behind him. Tentatively tapping Dr Ford on the shoulder, Taylor said: 'Pardon me, sir, but you did tell me to tell you. I believe you were talking to yourself.'

'Nonsense!' Henry snapped. 'I didn't hear anything!'

While it was true, as in this case, that he sometimes acted (brilliantly!) the role of an eccentric, it was also true that he really did talk to himself. John R.G. Turner, who was a graduate student in his department from 1962 to 1965, was rounding the end of a corridor one day, 'when Henry comes shooting out of his office, with something in his hand, saying to himself rather dramatically, "I think it could be done!" Then he saw me and looked embarrassed. So that was not for anyone's benefit.'

Henry loved to tell a certain after-dinner story. He and Sir Ronald Fisher had been doing a little amateur archeology in a churchyard in which there was one stone coffin with an open lid. They wondered if it was large enough to accommodate a human body, so Henry lowered himself into it. Just then, two old ladies arrived on the scene and peered into the coffin, in which the apparently lifeless figure of E.B. Ford was stretched out, eyes closed. 'Poor old gentleman,' one of the women was heard to murmur. Henry would always conclude: 'That was the most fatuous remark I ever heard!' His listeners couldn't help wondering if the two old women had not brilliantly turned the tables on E.B. Ford.

From the moment he began to set up his small Ecological Genetics kingdom, Henry started wooing Bernard Kettlewell. The moonlighting lepidopterist was well past his youth, and his lacklustre MA in zoology and medical degree were no kind of currency at Oxford, but Henry saw a greatness in this loud, eager man in Bermuda shorts. He knew that, although the clever genetics boys with their firsts might be good with chi squares, if an unusual moth or butterfly were to land in their trap they would not necessarily know whether its proper foodplant was comfrey, sallow or forget-me-not. Bernard would know this, and a thousand other things that are learned only by trial and error. He'd know what kind of container to put the pupae in, how far the males fly, how the insects 'overwinter' (as eggs, larvae or pupae), how long such and such a species remains *in copula*. He would figure out how to breed a male and a female in a small muslin cage suspended from a tree branch, and if that failed, he would manage to 'hand-pair' the insects, literally pushing them into the copulation position. He loved animals, was at home in marshes and fens, and would think nothing of jury-rigging generators in the pouring rain on some bleak northern isle.

'Bernard came from a very British background of fanatical amateur enthusiasm for natural history,' says David Lees, who worked as Bernard's postgraduate assistant in the 1960s. 'He always said we had to know about the "private lives" of the insects we were studying.' In the field he was said to be as lucky as he was skilled, or perhaps the two went hand in hand. On one occasion, finding himself surrounded by clouds of Bath white butterflies (*Pontia daplidice*) in South Cornwall, Bernard swung his net and caught a Bath white and a fabulously rare short-tailed blue (*Everes argiades*) at the same time.

It was not that Henry Ford did not know his Lepidoptera. He was the author of two classic books, one on British moths and one on British butterflies, which served to convert a generation or two of young boys to ecological genetics. ('I can date my professional life back to my ninth birthday when I got that book,' remembers Majerus. 'I'd been collecting bugs and this now gave me a reason for doing it. If I had to pick a scientific hero in my life it would be Ford. I am an ecological geneticist. He is the father of the field.') But Henry was first and foremost a geneticist. When an unsuspecting soul once came into the laboratories asking where he could find 'the great lepidopterist E.B. Ford', Henry replied: 'Never heard of him.'[40] He was not an outdoorsman of Bernard's calibre, not nearly so curious about insects' 'private lives'. In fact, he loathed all animals, with the exception of cats, which he doted on. He certainly had no affection for the woolly or twiglike caterpillars he brought home from the field, nor the butterflies that lay stunned in his bottles. Of a young technician in the department he wrote: 'She loves animals, god help her – imagine being fond of a caterpillar.'

For all these reasons Henry needed Bernard, and, surrounded by graduate students in their twenties, he doubtless longed for a friend closer to his own age. Beginning in 1948, he wrote letters in which he regretted that he had little money to offer but, like a bridegroom wooing a mail-order bride, lingered over the splendours of the nearby Radcliffe Science Library and promised to build Bernard a proper laboratory with electrical adding machines, laboratory assistants, and a new invertebrate breeding house. There would be a warm room with an incubator, and a refrigerator, and a veranda for hibernating insects, he added.

Bernard could have imagined no more glorious destiny than to be an Oxford scientist, but he was in no position to reincarnate himself overnight. Since 1948 he had been living in South Africa. What little passion he had for medicine had evaporated with the arrival of the British Public Health Service in 1946, and he had taken a job in locust control research at the University of Cape Town. In later years he would love to tell the story of how he inadvertently created a race of radioactive ants that escaped and set off scintillation counters all over the university.

How colourful, how sunny and temperate southern Africa must have seemed compared with the daily miseries, rations and fuel shortages of postwar England. In this scenic city stretched out around its deep blue bay, where bougainvillea, plumbago and pink oleander frothed over gleaming white walls, the Kettlewells – Bernard, Hazel, daughter Dawn, son David, and 'Nanny', Hazel's childhood nanny, who lived with the family and was adored by all – settled comfortably into the expatriate life. Half in jest, Bernard tried on the role of adventurer and Great White Hunter; he was a tolerable shot, always carried a rifle, and was inclined to pose, jocularly, in a Hemingwayesque attitude, chin thrown back, leg thrust forward as if he were standing on the corpse of a water buffalo. He loved tramping through an exotic continent brimming with tropical Lepidoptera, making unforgettable expeditions to the Kalahari, the Nyasa forest, the Belgian Congo, and Mozambique. On many occasions it would require all his fast-talking charm to dissuade border authorities from spraying his car and specimens with pesticides. His exuberant, outdoorsy ways, extravagant gestures and hard-drinking, hard-partying bonhomie were well suited to a colonial Africa approaching its twilight.

Hazel was very content in Cape Town, but she could sense that locust control in a provincial outpost would probably not satisfy her husband's ambitions. It was hard to refuse the call of Oxford, still less the opportunity to return to one's real passion on a full-time basis, and Bernard had always wanted his children to be educated in England. However, the Kettlewells were *bon vivants* with sumptuous tastes, and Oxford was offering a salary of only a few hundred pounds in 1949. After some of the astute politicking for which he was renowned, Henry managed to obtain grant money from the Nuffield Foundation, founded by the Oxford philanthropist Lord Nuffield, a.k.a. William Morris, the motorcar tycoon. Henry had a gift for conveying the sense that whatever he was doing at Oxford was of the most *urgent* importance; something in his tone suggested that the fate of nations might hang on ecological genetics. In 1951, he was able to cable Bernard: 'SUCCESS AT LAST, TRUST GIVES FULL GRANT.'[41]

This was the green light Bernard had been waiting for, but by now he was having second thoughts. In one of the glooms that periodically engulfed him, he had become morbidly preoccupied with the idea that a nuclear war with the Soviet Union was unavoidable. If so, wouldn't the children be safer in southern Africa? These thoughts paralysed his brain's decision-making centres for months, while Ford wrote consolingly that his diplomatic friends were of the opinion that there would probably only be 'a series of small wars', yet he didn't want to offer advice, he said, 'because had I done that, and you all came, I would never forgive myself if they were bombed.' Cockayne, meanwhile, was dourly predicting 'native troubles' in Africa and berating Bernard for deserting their joint collection and for being a fool. 'Either you . . . take up the Oxford job you were so eager to get, or you go into permanent exile . . .' At the end of the letter, he added drily: 'I got this started and then I got a heart attack . . . Most of my *repandata* [the butterfly *Cleora repandata*, or mottled beauty] were drowned in the sleeve.'[42] One problem was that if Bernard moved back to England he risked owing several years' back taxes, and to forestall this Henry and the Nuffield Foundation accountant talked the Inland Revenue authorities into a deal based on the national importance of Bernard's work. ('I don't see why you should get such special treatment from the income tax authorities,' the ever-supportive Cockayne wrote him gloomily. 'Your work is not quite so important to the country . . .') The deal was that for the first three years of Bernard's appointment as Senior Research Fellow he would divide his time between England and South Africa and would be spared British taxes.

At last, in September 1951, Bernard was ecstatic to find himself 'inhaling the aroma of guinea pigs and mice' at the School of Ecological Genetics. Still fearing thermonuclear war and in some doubt as to the solidity of his position, he had left his family behind in Cape Town for the time being. Henry had had a small garden dug next to the breeding house, planted out with larval foodplants: comfrey, horseradish, sallow bushes and dead-nettle. After colonial Cape Town, where some scientists espoused Lamarckian views and had never heard of industrial melanism, the intellectual air of Oxford was like pure oxygen to Bernard. He could discuss the delicious minutiae of Lepidoptera with his new lab-mate Philip Sheppard, for whom at first he served as a mentor, confiding all his knowhow about larval foodplants, hand-pairing, and such. The two men became fast friends. Everybody idolized the lanky, meticulous, acerbic Sheppard, with his keen intelligence and dark, penetrating eyes, even though his sarcasm could be mordant if you lacked intellectual ammunition. 'If you were going to tangle with Philip, you had better have your facts in order,' recalled one former student. Eventually, Philip would have to spend a good deal of time tutoring Bernard, who practically fainted at the sight of a column of numbers and was thrown into confusion by genetics.

In December Bernard was called home when Hazel was suddenly hospitalized for high blood pressure. He had been missing his family anyway, and even Henry, the quintessential bachelor don, could see that a Bernard abstracted out of family life was an untenable proposition. A modified tax deal of some sort was negotiated, to allow the Kettlewells to come to Oxford in the spring of 1952 and stay until autumn. However, Bernard didn't travel the easy way. His family went by ship, as one still did then, but he had always dreamed of braving the legendary Cape-to-Cairo journey in his beloved bottle-green Plymouth – something few, if any, people had accomplished, except perhaps in wartime, in jeeps or army trucks. Even today it is a desperate undertaking. Armed with four spare tyres, 24 gallons of petrol, jerrycans, chains, shovels, spades, a saw, ropes, spares, and enough Coca-Cola (which he drank instead of water, to the point of intoxication) to quench the thirst of a small nation, he barrelled across a now-defunct colonial geography – from the Union of South Africa to Southern Rhodesia, Northern Rhodesia, a large slice of Belgian Congo, and on to Uganda, tracing the northern edge of Lake Victoria to Kampala and eventually arriving in late-colonial Kenya. Along the way Bernard and his travelling companion, a photographer nicknamed Snap, contended heroically or hilariously with broken pistons, corrupt customs inspectors, malaria mosquitoes, tsetse flies bearing sleeping sickness, river crossings with the Plymouth floating on planks, Masai chieftains, pygmies, lepers, white hunters, Ugandan uranium prospectors, tomb robbers, elephants and rhinos. All of this is recorded in Bernard's highly readable travel diary, which was laboriously translated from his exuberant scrawl into type many years later by a heroically patient secretary.[43]

In a corner of the Belgian Congo the six-foot-four-inch Bernard walked into the rainforest with two guides and loomed, Gulliverlike, over 'a few 4-ft-high leaf covered huts, with these naked (almost) dwarfs cooking the most awful looking, filthy-smelling food in their pots ...' He had arrived at a pygmy village. 'We had previously bought 100 cigarettes (20 frs) and a log of salt which they dearly love ... We traded [them for] 2 bows and many arrows. The bows were afterwards tried out and they are immensely powerful and the arrows very sharp and thin.' The next day, upon crossing the border into Uganda, a sudden tropical downpour turned the road into 'deeply rutted slime', in which the Plymouth slid sideways without warning and became mired. Near the notice 'Elephants have priority in crossing road – on no account leave car' they had to be pushed out 'by natives'. A few days later the diary records that the photographer was in his 'worst mood of bloodymindedness' yet and a parting of the ways occurred abruptly afterwards. In Nairobi, where the imperial lifestyle was obviously on its last wobbly legs and 'the Europeans talked openly of what they expect to happen', Bernard was forced to concede that the Plymouth was not going to make it to Cairo. After arranging to ship it to England, he proceeded north by train, boat and truck. In Sudan he shot a crocodile with his host's Remington rifle, and then strolled up to a native market 'and saw more bad fish and nakedness'. In Egypt 'dancing girls were brought before us and performed the "shimmy" in no restrained manner. [The tour guide] explained the girl was ours either on a boat up the Nile or at her house . . . She would even wear men's clothes if we liked!' He declined the offer according to his diary account.

Eventually, he made it all the way to Oxford, within whose sombre stone walls Bernard put thoughts of harem girls and crocodiles behind him and focused on the plan incubating in his mind. It involved *Biston betularia*. He would find out if birds really did eat more black moths in the unspoiled woods with lichened tree trunks, and more pale ones in darkened, polluted places. Using the 'mark–release–recapture' technique pioneered by Fisher and Ford in their early fieldwork, moths could be marked with a dot of enamel paint and released. The position and colour of the dot functioned as a dating system, so that when recaptured at a light trap, an insect's vital statistics could be recorded. Originally, the proportion of marked insects recaptured was used to estimate the entire population of a colony insect. As a result of Fisher's genius, the numbers were arrayed in a triangular trellis-like chart conveying arithmetically the salient facts of population size, births and deaths. Fisher had always foreseen that his method could be extended to estimate the differential elimination of different phenotypes, as Kettlewell would use it for the peppered moth. He would conduct his experiment in two places: a polluted woodland and a pristine rural environment. If birds were preferentially eating moths that stood out against their background, there should be a difference in the recapture rates.

The experiment he imagined would require almost as much planning as a major military campaign. He would need many hundreds, even thousands, of peppered moths to mark and release, and all of these had to be bred, and they would have to emerge at the proper time. Bernard had started breeding peppered moths the previous summer – he needed these for his ongoing breeding experiments, for he was trying to pry apart the genetics of the species – and had packed away some pupae to overwinter in the Insect House. But most had died and the survivors were mouldering and he cursed himself for leaving them too wet. Only because a few collector friends answered his frantic SOSs and sent him pupae was he able to go forward with his breeding experiments.

The moths he would mark and release the following summer of 1953 in his experiment had to be bred from the 1952 crop of twiglike caterpillars, ranging in colour from green to purplish-black. Accordingly he was rearing countless thousands of hungry *Biston* larvae at Wytham Wood, an unspoiled university-owned woodland west of the Thames, in large muslin-wrapped tubs and in muslin sleeves tied around the branches of sallow bushes. The sallow, which has white flowers with a sickly sweet smell, is *Biston*'s major foodplant. Bernard had to camp there full-time, while his family stayed with Hazel's mother in Surrey, for he was compelled to work almost around the clock defending the sleeved caterpillars against predators and running the mercury vapour light trap at night to tally the numbers of typicals and *carbonaria* in the area.

At summer's end a large trailer parked in Wytham Woods became the Kettlewells' home in England (there were undoubtedly tax reasons behind this, too). The children were already accustomed to this way of life, falling asleep in the woods to the steady hum of generators and the enchanted glow of mercury vapour lights, watching their father get worked up about things with latinate names. Nine-year-old David was himself becoming keen on entomology, and Bernard was starting to take him along on his visits to Cockayne to work on their collection. (To the boy the aloof, hawk-nosed 'Cocky' seemed a character out of Dickens, and a visit to his musty house, where the windows were always sealed, the curtains always drawn and no meal ever served, was a thing to be dreaded.) Dawn, a wilful fifteen-year-old, was more interested in boys than moths, so much so that her parents were becoming alarmed. Both David and Dawn liked coming to visit the Oxford genetics laboratories, where the abundance of mice, guinea pigs, fish and birds almost made up for the fact their father had so little time for them.

In early September, the whole family performed an emergency evacuation of the precious peppered moth caterpillars, which were under attack by shrews, wasps and other predators. Nanny, who sewed all Bernard's muslin sleeves, was mending holes as fast as the wasps could drill them. Bernard always smoked cigars incessantly, drank and ate to excess, and already suffered from colitis as well as recurrent attacks of bronchitis, influenza, and other 'hideous plagues' whose symptoms he was prone to describe in the laborious detail of the medically-trained. He kept irregular hours, lost himself in long periods of overwork and no sleep, and had tempestuous mood swings. In short, he was frequently a wreck. Nonetheless, by mid-September, when his caterpillars were ready to pupate, he managed to pack away three thousand pupae in film cans.[44] They would slumber through the winter in the Oxford School of Genetics' 'Insect House' on blotting paper atop slightly damp peat moss, and when they emerged as moths next June they would gain eternal fame – that is, if all went well.

Beneath his boisterous exterior, Bernard was feeling jittery. He had left secure employment to stake his entire career on a moth about which he actually knew very little. (He had dashed off a note to a collector friend, a retired colonel, asking nervously to know 'exactly how you winter this species. Are they in fact kept bone dry or slightly damp or on damp fibre, and whether you keep your tins in an outhouse or indoors.')[45] The mark–release–recapture method he would use required that a fair percentage of the marked moths should *return* to the traps to be counted, but *Betularia* were known as strong flyers, unlike some sedentary colonial moths. Perhaps the moths would simply fly away. Maybe he had chosen the wrong species. 'My worry is that, whilst providing excellent material in every other way, I feel it may turn out to be a great flyer so that only a small proportion of "returns" will be collected,' he wrote to the colonel.[46] Having picked on a possibly fugitive moth was, however, only half of the problem, as Bernard was about to discover.

5

E.B. Ford and his enemies

Bernard's moths had much to prove. The redoubtable Henry Ford had an agenda, and everyone who worked in his shop absorbed it from the air. There were certain terms that could not be uttered in his presence. Certain books or papers, written by certain scientists he considered his rivals, were the works of the Antichrist. When Julian Huxley appeared in the laboratory one day singing the praises of a book by a scientist Ford deplored, he was heard to ululate: 'One should not judge a book by its avoirdupois.'

'I remember once mentioning random drift to him,' said Lincoln Brower. 'That was not a good idea.'

Random drift was an anathema Henry resisted all his life. When a population is small and isolated, the American population geneticist Sewall Wright proposed, certain genes might shift in frequency purely due to chance, or 'sampling error', and this could become a force of evolution. Sampling error might not seem like much, but it could accumulate exponentially like compound interest. Suppose you have isolated a village of a few hundred people, of whom, say, seven adults are red-haired. If ten of the village's black-haired young men happen to fall off a cliff one day during a hunting expedition, this accident will leave the redheads to pass on a higher percentage of the next generation's genes, and the shift in gene frequency will become more pronounced in the subsequent generation, as each red-haired child, now grown up, produces, say, three or four more redheads. Yet red hair confers no fitness advantage per se, no effect on survival or fertility; the gene's ascendancy would be entirely due to chance.

Dramatic examples have been documented among isolated population groups, such as certain South Sea islanders and the 8,000 Amish of Lancaster County, Pennsylvania, who are descended from a mere 100 'founders'. The latter suffer from a high rate of genetic diseases including pyruvate kinase (PK) deficiency, a form of hemolytic anemia, traceable to a single eighteenth-century forefather, 'Strong Jacob' Yoder. This was a striking case of the 'founder effect', a form of drift. Here a gene with no selective advantage – indeed quite the opposite – had spread from a single 'founder' individual to much of the population. The fact that speciation, the formation of new species, was believed to arise under just such circumstances, in small, geographically isolated populations, gave the question a special urgency. Was one to believe that the evolution of species was essentially random?

Julian Huxley always referred to random drift as 'the Sewall Wright effect', and Ford and Fisher abhorred it with the same vehemence that the medieval Church Fathers abhorred the Cathar heresy. Mutation was random, of course, but evolution should not be. 'Evolution is progressive adaptation and nothing else,' Fisher had written. It was common in those days to speak of natural selection as a progressive 'force' that 'acts on' or 'moulds' or 'sculpts' an organism. To some extent this was just a convenient shorthand, but it also reflected the notion that a given feature is expressly designed *for* some purpose: that the ample whiskers of such and such an animal evolved to enable it to gauge the width of caves, or tickle its mate, or whatever. Imbued with teleology, natural selection becomes a purposeful 'agent' or 'force' that is always tinkering with creatures, perfecting them, not unlike the old clockmaker deity, and quite clearly Fisher's thinking was tinged with this view. As the evolutionary biologist John Endler puts it in *Natural Selection in the Wild*, 'Natural selection no more "acts" on organisms than erosion "acts" on a hillside. It is a result of heritable biological differences among individuals, just as erosion is a result of variation in resistance to weathering and running water.'[47]

During the 1940s Henry was fighting battles left over from the 'eclipse of Darwinism' period. Back in 1926, the authors of an influential book, O.W. Richards and G.C. Robson, had examined every single putative case of natural selection in the wild (house sparrows, crabs, snails, butterflies) and pronounced that not a single one was convincing. It was highly unlikely, they said, that natural selection played any role in shaping taxonomic differences, the visible characters used to distinguish among closely related species and varieties; these had evolved at random. The notion was accepted by almost everyone at the time, inspiring Haldane, who was a great believer in selection, to lament: 'This has led many able zoologists and botanists to give up Darwinism.' At the same time the British Museum's chief ornithologist closely examined Darwin's collection of Galapagos finches, the most sacred of Darwinian icons, and announced that there was 'no scope for selection' in this 'heterogeneous swarm' of birds.

In the 1930s even Ford admitted that non-adaptive characters existed, but he insisted that the genes responsible for them were probably also doing other things that *were* adaptive and thus were subject to selection.[48] So convinced was Ford by Fisher's genetic determination, that on the basis of scant evidence he was prepared to attribute unspecified and hidden adaptive functions to the nonadaptive genes.

The battle between the forces of randomness and the forces of selection came to centre on a certain day-flying moth, the scarlet tiger, *Panaxia (Callimorpha) dominula*. The normal genotype is almost tropical-looking, with scarlet hindwings, but there were two other types. Henry had first run across the rare *medionigra* type in a colony in Cothill Marsh near Oxford in the late 1930s, and found few like it in old collections in museums and private houses. This might be the long-awaited miracle: a new gene just beginning to spread in an isolated colony by means of natural selection. And, fortunately, it was a made-to-order conspicuous polymorphism, controlled by simple Mendelian inheritance. If he returned year after year, Henry thought he could record the change in gene frequencies and calculate how great the selective pressures were. He rushed to inform Fisher, who exulted: 'This is a very ripe plum waiting to drop in our mouths!'

Ford and Fisher believed the scarlet tiger could defeat random drift. Although Wright had never claimed that random drift accounted for *all* genetic change, Fisher and Ford always acted as if it was all or nothing. The Cothill colony, in their view, was a population small and isolated enough to be a worst case for *their* theory (because this was just where random drift was supposed to wield the greatest effects), so they could say to Wright: 'Here is just the sort of place where your drift should flourish, and look – it's all natural selection!'

Henry began his annual *Panaxia* census in 1941, accompanied by the short-sighted, absent-minded Fisher, then in his fiftieth year, who had to be reminded to wear suitable clothes and Wellington boots. The census went on for years – indeed it continues today, long after Ford's death – and every year Henry would await the figures like a nervous financier poised to hear the new prime lending rate. The percentage of *medionigra* fluctuated a good deal from year to year – from 20.5 per cent in 1940 to 7.9 per cent in 1946 – and the population itself climbed from about 1,000 adults to around 7,000. Ford and Fisher read great significance into these figures, which they claimed, in a famous 1947 paper, were 'fatal' to Wright's theory. The robust fluctuations in gene ratios at Cothill showed that *all* populations, large or small, were subject to 'selective intensities' too large to be ascribed to drift. The arrogance of the language was striking, and typical of both scientists' habit of thinking in absolutes. It did not concern them that no one knew what precisely the selective advantage might consist of, how one or another genotype might be better adapted to its environment.

Wright penned a heated counterattack, which triggered a further fusillade from Fisher and Ford, most of the bickering occurring within the journal *Heredity* in the years 1947–51. The feud provided ready entertainment at scientific meetings. According to William Provine: 'Everyone used to seat Fisher and Wright next to each other to see the fireworks. They would turn their backs to each other and never speak.' In 1949, Wright came to Oxford, where, speaking in his trademark Midwestern monotone, he gave a badly organized two-hour lecture illustrated with slides that no one could discern. After the decorous applause, Ford declared that he was *entirely in agreement* with Dr Wright, with a few unimportant exceptions, and then proceeded to take him apart for a quarter of an hour, concluding that 'with these trivial exceptions it is a pleasure to express our complete agreement with Mr Wright.' Wright said something to Fisher, who, seated stonily in the front row with his arms folded across his chest, finally barked: 'I only went so far as to disagree with you completely. Could I do more than that?' Later the British grapevine reported that Wright had been 'annihilated'.[49]

But there really was no definitive winner, and Ford knew that the random drift heresy was still widespread, especially in America. Throughout his life, with assorted underlings in tow, he would tramp through the Cothill marshes every year in late June or early July, when *Panaxia* is on the wing, to get the counts. He was also on the lookout for other species that would prove his point. He badly needed to find some really powerful cases of selection in nature.

Snails came to the rescue. While Kettlewell was still irradiating locusts in Cape Town, Ford wrote him excitedly that 'the snail genetics is turning out most exciting'.[50] One day Philip Sheppard, the department's golden boy, had dropped by the college rooms of his friend Arthur Cain, who was a lecturer in Zoological Taxonomy. Cain took down a sack and poured out a cascade of shells of *Cepaea nemoralis*, a common European land snail. As a lepidopterist, Sheppard knew next to nothing about snails, but he could see instantly that the differences in the shell patterns were striking: Here were shells with bands and shells without, yellow shells, brown shells, pinkish shells. Such variations *couldn't* be random, Cain and Sheppard felt sure; they must mean something. Yet *Cepaea nemoralis* had figured in Robson and Richards' book as an exemplar of a conspicuous polymorphism with no adaptive value.

In an elegant piece of fieldwork, Cain and Sheppard proceeded to demonstrate that the different forms of *Cepaea* corresponded to the animals' background. In England the unbanded brown and unbanded red forms were more common in dense woods littered with leaves, while the yellow-green forms and the banded forms seemed to be adapted to lighter backgrounds, such as grassy areas and the green foliage of hedgerows. Since the snails were eaten by song thrushes (which used rocks as anvils) Cain and Sheppard could count broken shells at the rock site and compare them with their census of live snails in the surrounding area. Thus they could determine the intensity of predation. Banded forms crawling around on a uniform background, where they were more conspicuous than their unbanded brethren, were preferentially eaten by the birds, they determined. Conversely, unbanded forms were preferentially gobbled up on rough or disturbed backgrounds. Each colour and banding type seemed effectively camouflaged in the proper setting.

It certainly looked like a showcase example of natural selection, and it was all the more exquisite for having been singled out previously as an example of a *nonadaptive* polymorphism. Cain and Sheppard were quickly becoming heroes of the adaptationist cause, and *Cepaea* would soon be a superstar in all the textbooks. 'It seems that in view of the results . . . all such cases of polymorphic species showing apparently random variation should be reinvestigated . . .' Cain and Sheppard concluded in their 1950 paper.[51]

Just because you can't see the adaptive purpose of a given trait doesn't mean it doesn't have one, Cain would always insist; it simply means you need to look harder. Now it seemed that 'very considerable selection coefficients (even up to 50 per cent in some combinations of genes investigated by Fisher) are actually found to act in the wild . . .' he said. Moreover, as Ford had predicted over a decade earlier, 'even apparently trivial characters are far more likely to have their frequency and distribution determined by selection than by random effects . . .'[52]

At stake was nothing less than the mechanism of evolution. By the early 1950s all of the leaders of evolutionary science had come around to the view that most conspicuous polymorphisms were the handiwork of selection, not random drift. This shift in *Weltanschauung* owed much to fieldwork demonstrating that the three most frequently cited examples of nonadaptative differentiation – the *Drosophila* inversions studied by Dobzhansky, banding patterns in *Cepaea*, and blood groups in humans – were all subject to strong selection pressures. Addressing the Golden Jubilee of Genetics in Columbus, Ohio, in 1950, Julian Huxley exulted about the 'all-pervading influence of natural selection, and the consequent omnipresence of adaptation', and pronounced random drift a rare occurrence.

Just about everyone – certainly everyone in hyperselectionist Britain – now agreed. 'In England, or at least within the widespread influence of the Oxford school of ecological genetics,' according to John R.G. Turner, 'a belief in strong selection coefficients and the totally inefficacy of random drift became *de rigueur*. Deviation within the group was not permitted. Outsiders who came up with interpretations of field phenomena involving significant amounts of genetic drift . . . got some pretty rough handling.'[53]

Yet to establish that selection operates very strongly in some instances does not mean that it is all-pervasive in nature. As critics would point out a few years later, E.B. Ford and his circle were actively – and hence preferentially – searching for examples that would prove their case.

By the time Bernard Kettlewell returned from South Africa, Henry Ford had reason to gloat. *Cepaea* was everything he had ever hoped for; the forces of random drift were in retreat; his protégés, and especially Philip Sheppard, were doing exemplary work along the lines he had always imagined. The Oxford

School of Ecological Genetics was falling into a predictable set of rhythms.

Every year there were the midsummer butterfly rituals. *Panaxia* emerged in late June or early July – depending on weather conditions – and for two weeks everyone in the department took turns chasing tiger moths at Cothill and recording the phenotypes on specially printed score sheets. After a long day wading through the marshes, as the fairy pinks and blues of the northern midsummer skies faded to deep indigo, it was traditional for the returning Cothillers to swing from pub to pub in downtown Oxford. Kate Coldham Davies, who spent four years as department secretary, observed Henry saying to his troops: 'All right, we're going to try all the beers in Oxford.' Because he would always insist they have a gin to 'clean the palate' before every beer, after nine or ten pubs 'everyone would be reeling, but Henry would never show any signs of drunkenness,' said Davies. On one such evening, the Oxford geneticists ran into a woman collecting for the National Society for the Protection of Children. Ford rushed up to her, proffering a five-pound note, and said, in his most unctuous manner: 'I am soooo glad someone is doing something to prevent children.'

Around the time the Cothill rites ended, the 'jurting' season began. Successive generations of 'jurt boys', or 'jurtinologists', as Henry dubbed them, would journey with Henry to four or five familiar fields on the Cornwall–Devon border, to catch meadow brown butterflies, *Maniola jurtina*. At day's end they always repaired to their favourite inn, and after dinner everyone would sit on someone's bed and count the spots on the wings of the dead butterflies that made up that day's catch. The different spot patterns were a genetic polymorphism that Henry had been tracking since 1940, following with intense interest the changes in frequency from place to place and year to year. He had also been keeping a record of the spot patterns of the meadow brown in the Scilly Islands, off the coast of Cornwall, particularly a population on the small uninhabited island of Tean. In 1951, a newspaper article appeared portraying the butterfly-counting E.B. Ford as a 'modern-day Crusoe'. 'Living under canvas with a colleague, Dr E.B. Ford, F.R.S., will be cut off from all other human contact during his stay,' the article reported. 'The island, one of the Scilly group, is "secret".' The secret part was Henryesque. 'It is most essential that we should not be disturbed,' he was quoted as saying. 'A moment's interruption could cost me three hours' work.' Apparently unable to grasp exactly why the professor was doing this, the reporter concluded that his methods could be applied to counting rats, and hence to preventing future outbreaks of bubonic plague.

When he was not in the field, Henry, always exceedingly punctual and self-disciplined, would arrive at the laboratories at nine o'clock every morning. On fine days he rode his bicycle, his pressed trousers secured by bicycle clips, peering at the bustling traffic of North Oxford through his heavy glasses. On other days he would take the bus down the Banbury Road from the pleasant fawn-coloured stucco house he shared with a housekeeper on Apsley Road in tree-lined North Oxford. He was always formally dressed in a dark suit and tie; his brown brogues always spit-polished.

From his office radiated the mood of the day, as a low-pressure front over Newfoundland affects the skies of Boston. The walls were paper-thin, so no one's state of mind was really a secret. When he wished to ask colleagues or employees how they were, Henry invariably asked: 'And how are your circumstances?' prolonging the vowel of the penultimate syllable. His entourage adopted his terminology. 'And how are Henry's circumstances?' they would ask one another. If Henry's circumstances were good, everyone knew it because he would say 'Glop!' as he trotted down the corridor. If they were not so good, he would say something like: 'The day is like a bad pig!' (The many things he disliked were 'like a bad pig'; nobody knew where he picked up the expression.)

Towards his secretaries he was typically gracious, even courtly, gratefully acknowledging their typing and research help on his books. He always doffed his hat to women. According to Ruth Wickett, who worked for a while as his secretary, 'He was nice to me because he needed my help. But towards other people he could be horribly rude.' One wife of a graduate student, who worked part-time as the departmental librarian, once came to his office to collect an overdue book. 'Come in,' said Ford when she knocked. When she entered the room, he immediately started dictating a letter.

'Dr Ford, I am *not* your secretary.'

'Then go away.'

Members of his department vividly recall dinners at 5 Apsley Road, where Henry was a cordial host who served sherry in antique glasses with long, twisted stems in which bubbles were suspended. This collection of stemware was one of Henry's prides, about which he would discourse at length. When Merrill and Laurence Cook were entertained by Ford at All Souls College shortly after becoming engaged, Merrill remembers: 'He brought out these sherry glasses that were obviously exquisite, and said to me, "I do-o-o think the enamel stem period is the most beautiful, don't you?"' He had the refined and recherché tastes of an effete caliph. He collected liqueurs. He had never met a heraldic crest he could not decode – a habit that Bernard liked to josh him about ('Oh Henry, how are your rabbit feces rampant?'). He owned several John Ruskins. He did not own a television or radio and never read a newspaper, but he always knew what was going on in the world. (In his later years a television entered his house because his new housekeeper was addicted to soap operas, but he would not allow it to be turned on in his presence.) It was inferred, probably correctly, that his politics were derived from dinners with ministers, heads of state, and shadowy figures from MI5. These were the sorts of people with whom he was said to share sherry at his rarefied London clubs and later at All Souls College in Oxford.

His domestic arrangements included a housekeeper, without whom he was rather helpless, and Kevin,[54] a muscle-bound young man who was treated like an adopted son, sent off on travel holidays, and whose provenance was never explained. Ford was sometimes brusque and bullying to Kevin in front of guests, criticizing him about his table manners and lack of savoir-faire. When the wife of a graduate student once ventured to ask: 'Dr Ford, how did you come by your ward?' he answered, enigmatically: 'How *does* one come by one's ward?' A friend from the time had no doubts about Henry's sexual orientation: 'Of *course* he had boyfriends. But it was very dangerous. You could be thrown in gaol like Oscar Wilde.'

'Everyone assumed he was wildly homosexual,' says Laurence Cook, who came to the department as a graduate student in 1957. 'He used to go off on these long weekends in Amsterdam, and everyone said: "Ah well, of course." As far as I know there was no evidence whatsoever.' Once young David Kettlewell happened to bump into Henry on an Amsterdam-to-London flight. The next day Bernard said to his boss: 'Oh, David mentioned he saw you coming back from Amsterdam.' Henry replied: 'I didn't see him. He must have been mistaken.' Some people insisted he was not homosexual at all and that he had always been tragically in love with his cousin and dear friend Evelyn Clarke. If there is a grain of truth to this, perhaps it explains his spirited defence of 'inbreeding' in at least one genetical monograph, in which he objected to the Roman

Catholic ban on the marriage of close relatives and claimed that 'wide outcrossing' was worse, genetically speaking.

It was a matter of record that Henry counted many famous people among his friends and that he was apt to remark that a page or two of Debrett made comfortable bedtime reading. An aristocrat manqué, he would become giddy in the company of actual aristocrats. 'On two occasions when he came to visit me the Duchess of Devonshire happened to be there,' Rothschild recalls, 'and he was overjoyed. Another time Diana Cooper [the author and society figure] was visiting. Ford was beside himself with pleasure. He would become very animated, telling amusing stories, overacting and flattering them too much. He had the *folie de la grandeur*.'

Whatever he lacked in social pedigree he more than compensated for with his regal command of his Oxford subdepartment. Henry had always played up his role as the saviour of Darwinism during the beleaguered 'eclipse' period, all but claiming blood ties to the Darwin family and magnifying what appear to have been a few offhand conversations with Darwin's son Leonard into an intimate lifelong association. 'It is just what Father would have so much wanted someone to do,' he pointedly recalled Leonard Darwin telling him about his research. With a compulsion bordering on the fetishistic, he would periodically invoke this direct Darwinian transmission to bestow legitimacy on his Ecological Genetics department, which sadly lacked material existence in contrast to the bureaucratically real subdepartments directed by David Lack (the Edward Grey Institute of Field Ornithology) and Niko Tinbergen (Ethology). By and large, he seems to have come to believe his own public relations, and created a general impression that the spots on *Maniola* butterflies and the annual counts of *Panaxia* mutants were changing the course of evolutionary history. Younger scientists in his orbit would look back on the period all their lives as a brief, radiant Camelot, bathed in the golden beams of Ford's peerless intellect. Miriam Rothschild, now in her nineties, savours her memories of the 'extraordinary stimulating atmosphere of that time – it sparkled!' He had a great gift for inspiring junior researchers in his sphere of influence to dedicate themselves wholeheartedly to his great cause, and H.B.D. Kettlewell, lately a general practitioner and anesthesiologist, was as enthusiastic and devoted a disciple as any. When Bernard went into the woods with his peppered moths, he would have fervently hoped to find support for his boss's theories.

As Henry's disciple, Bernard Kettlewell was in a direct line of descent that traced back to Darwin, and the experiment he was undertaking would be the fulfilment of a sort of scientific grail quest. The population geneticists had elaborated theories about how evolution worked, how much variation was heritable, how much of it was due to natural selection and adaptation and how much to chance, but no one really knew how relevant these models were to what was happening in real populations. In the wars of selection versus chance, adaptive versus nonadaptive change, Fisher, Ford, and the 'Oxford School' were largely successful in vanquishing random drift and nonadaptive change and making the world – the Anglo-American world, anyway – safe for strong selectionism and the belief that most traits had adaptive value if you just looked hard enough to find them.

The notion of adaptation, the fit between organisms and their environment, seemed intuitively obvious yet fiendishly tricky to prove empirically, given the long time-periods involved. It was a corollary of Darwin's theory that the dramas of adaptation and natural selection should be revealed best in times of rapid environmental change, when organisms are under pressure to adapt quickly to a new set of circumstances, and indeed the Industrial Revolution in Britain produced the greatest man-made environmental change yet observed. As trees turned dark and the lichens disappeared, a startling alteration in a species took place. It was a rare and blessed event to find a polymorphic trait apparently so crucial to survival as the wing colour of the peppered moth, and Bernard was determined to satisfy his boss and Oxford that he could document and measure natural selection in action.

PART II

6

Triumph in Birmingham

On the second day of June 1953, most Britons were huddled in front of the small black-and-white television sets of friends or neighbours watching the coronation of the twenty-five-year-old Queen Elizabeth II in Westminster Abbey, the first state occasion to be televised in Britain. Not long afterwards striking panoramic photos taken from Mount Everest by Edmund Hillary and Tenzing Norgay, who had summited at 11.30 a.m. on 29 May 1953, were showing readers of *The Times* what the icy 'Roof of the World' looked like. Three days after the summer solstice, Henry Bernard Davis Kettlewell set off for Beacon Wood, near Birmingham, to conquer a small but mysterious precinct of nature.

After some morning thunderstorms, it was a fine day, humid and warm. Bernard was at the wheel of his much-dented

Plymouth, which had survived being driven from Cape Town to Nairobi, and which had since sent many a pedestrian and bicyclist on English lanes scrambling for cover, for Bernard was a notoriously exuberant and reckless driver. On that day both car and trailer were laden with mercury vapour light traps, 'assembling' traps, *betularia* pupae curled up in their sleeves, assorted tools, camping gear, food and water, not to mention enough cigars and gin to last until his 'return to civilization', as he put it. Because peppered moths emerge from their cocoons during a period of three weeks in late June and early July, Bernard had only a narrow time window in which to carry out his experiment.

His mission was to demonstrate experimentally that the darkened (*carbonaria*) form of the peppered moth survived better in an industrial area, and that this was the result of superior camouflage. Using Ford and Fisher's mark–release–recapture method, he would release paint-marked moths every day and find out if he recaptured more *carbonaria* than *typica*. The rate of recaptures would have to be significantly different; there would have to be adequate controls; and Bernard would need to show that the camouflage worked by protecting the moths from bird predators.

Bernard wasn't exactly in the wilderness. Beacon Wood, also known as the Cadbury Bird Reserve, was on the estate of his old friends, Christopher and Honor Cadbury, members of the Cadbury family that made its fortune from chocolate. It lay six miles from the grimy heart of industrial Birmingham, and years of heavy pollution had stripped all the trunks and boughs of lichen. Bernard's outdoor workshop was a triangular slice of woodland lying in a hollow between two hills, dense with oaks and a sprinkling of birches, with a thick bracken undergrowth. Fields and gardens bordered the experimental area on three sides, and the fourth side merged into the main part of the woods.

Bernard began unpacking the gauze sleeves containing three thousand pupae on the verge of hatching and hung them from tree branches near his caravan, where they could be cared for and inspected several times daily. Their contents had been lovingly nurtured since the end of the previous summer from tiny larvae that grew, in successive instars, or moults, into larger caterpillars, then wove the loose cocoons wherein their pupae safely overwintered at Oxford and survived the various catastrophes that might otherwise have befallen them. Their emergence had to be carefully orchestrated – sometimes Bernard slowed the pupae down in the lab fridge or hastened their progress by 'forcing' them, warming them up in the incubator – for these were meant to be the moths that proved Darwin right.

Under Bernard's large, capable hands an outdoor laboratory took shape. Extension cords snaked to two large mercury vapour light traps (which capture moths attracted by the glow). Around the perimeter of the area hung a number of home-crafted 'assembling' traps of muslin and perforated zinc, each containing two virgin female moths (one of each phenotype). The females would 'call' with their musky phero-mones, and as males 'assembled' to couple with them, they would become trapped. A male moth, wrote E.B. Ford in his classic book *Moths*, 'will fly from the country into town, with its manifold smells, in order to reach a female and into a room filled with the reek of smoke'. Both types of trap were destined to recapture some of the lab-bred moths that Bernard would mark and release every day; they would also provide more wild-caught local moths for the experiment. Bernard also hoped that by flooding the woods with sexual scents from the assembling traps he could keep enough marked and released moths in the vicinity. Knowing that *Biston* were strong flyers – they migrated – he was afraid that if too many flew away his experiment would be ruined.

In the lingering northern twilight, Bernard would have felt rather like a general on the eve of battle. All his entomological campaigns were life-or-death affairs in his mind, and this one was even more fraught, because it carried the weight of evolution and Oxford and E.B. Ford. It was Bernard's style to become quite manic in the field – sleeping little, forgetting to eat, frenetically overdoing, shedding alarming amounts of weight in a few days – and on this occasion he would have been reeling from the complex preparations, the drive, the heavy lifting, and his ever-present anxiety. His lumbago was acting up, and as the pale sherbet colours overhead darkened to violet and indigo, several gins and a succession of cheroots would have been required to soothe his frayed nerves. Perhaps, as he checked on his mercury vapour lights, he looked up to see Saturn perched in the southwest quadrant of the night sky above the bright star Spica, Arcturus conspicuous near the dome of the midheaven, and the constellation Leo slowly setting in the west. In the morning he would find out what moths he had attracted.

He had done a preliminary experiment a few weeks earlier to establish two crucial facts: that birds *did* eat *Biston betularia*; and that crypsis protected the moths from such predation. With his moths in the back seat, he drove to the Madingley Research Station in Cambridge and, with a colleague, set up a predation experiment with two great tits in a cage. If the birds ate the moths at all, he needed to find out if they would pass up the better-camouflaged insects in favour of the conspicuous ones. Posing the peppered moths on an assortment of dark and light 'furniture' trunks and boughs, he let the great tits in to the buffet. For two hours they paid no attention to the moths. An hour later they ate up all the conspicuous moths but also two camouflaged ones, leaving only three moths untouched.

The next day the birds devoured sixteen moths in half an hour, leaving only two survivors. Bernard concluded that 'the tits were becoming specialists on betularia and subsequently they were seen to be searching each tree trunk eagerly one at a time, immediately after admission, thereby defeating the object of the experiment.' All he had created was an elaborate bird feeder. Things were going badly.

The next day, probably in some desperation, Bernard tossed in a smattering of other local moths and insects along with the peppered moths, and this stratagem worked. All eight conspicuous moths were eaten and only half of the inconspicuous moths. Through field glasses, Bernard tried to keep track of the order in which the birds ate the *betularia*, but they worked too fast. He would claim in his paper, written in 1955, that the aviary experiment proved that the great tits did eat stationary *betularia*, and that 'they took [the moths] in the order of conspicuousness similar to that gauged by the human eye'. This may have been true, but only on the third or fourth try, and under rather artificial conditions.

At the aviary he and his co-workers had devised a method of rating a moth's degree of crypsis against a certain background. They would pace out the number of yards they had to walk away before the insect merged into its background. On a six-point scale, from -3 to +3, anything visible from 30 yards away was given a score of -3 ('very conspicuous'), while an insect invisible at two yards was scored +3 ('very inconspicuous').[55]

In the woods near Birmingham, Bernard would no longer bother to pace the distance but would simply eyeball the moths to assign them a score. He began releasing his first batch on the morning of Thursday, 25 June, on which day *The Times* was bustling with news of the second successful British atomic test, in southern Australia, and of the Everest heroes then toasting their triumph in Calcutta – not that Bernard had time for the news. 'Releasing' does not quite convey what he did. As each moth emerged from its pupa, he marked its underside with a dot of quick-drying enamel paint, put it in a small box, and then shook it gently onto one of the 33 boughs and trunks he had selected and numbered. (The moth would soon be joined by four or five others on the trunk.) Peppered moths, being nocturnal, are in a cataleptic trance in the daytime, and would remain motionless for the rest of the day wherever Bernard placed them. He made a point, however, of describing how the insects wandered about a bit on the trunk before settling down, as if searching out the perfect resting site, an idea that would come back to haunt him. He scored each moth on his six-point crypsis scale – after making sure that the sun did not shine directly on the insect, which would have fried it – and went on to the next one.

The ridge of high pressure covering the British Isles remained on Friday, bringing more fair, warm weather that ought to have been good for collecting, but Bernard found only five moths with dots in his traps that morning. A poor catch. The next day, Saturday, brought just two recaptures. There were four on Sunday, 28 June, nine on Monday, 29 June, and on the morning of 30 June, when in the outside world the Henley Regatta was in full swing with all its pageantry, only two marked moths waited in his traps. It was the very disaster Bernard had dreaded the previous summer when he thought about switching from *betularia* to *Gonodontis bidentata* moths, which migrate less. His recaptures were too low! Even with his modest grasp of statistics he realized that recapturing one black moth and one white moth, say, or two black moths and one white moth, out of 59 or 63 moths released, would not be statistically significant. If he couldn't prevent his moths from flying away, his experiment would be meaningless.

We don't know exactly what state Bernard was in, but we

can deduce something of it from a letter dated 1 July from Henry, who wrote: 'It is disappointing that the recoveries are not better . . . However, I do not doubt that the results will be very well worth while . . .'[56] The message sounds benign enough, but, knowing Henry, Bernard might have decoded it to mean: 'Now I do hope you will get hold of yourself and deliver up some decent numbers.' At the best of times Bernard was an excitable man, with moods prone to rollercoaster from giddy highs to Stygian lows. He now felt that his fate as a scientist was hanging on this experiment. The stresses of his lumbago and the difficult fieldwork, not to mention the pervasive pressures of Oxford, would not have brightened his outlook.

In the paper Bernard would publish on this experiment there is a table, Table 5, listing releases, catches and recaptures for each day. Squinting at the columns of numbers, we notice a strange thing: from 1 July on, after the letter from his boss, the recaptures suddenly soar. After recaptures running in the low single digits for the first six days, on the morning of Wednesday, 1 July, Bernard found 23 moths with paint marks in his traps, 19 of which were dark and 2 light. (He set out and recaptured the intermediate *insularia* form as well but for the sake of simplicity we will ignore these.) Thursday, 2 July, brought 34 recaptures, including 28 darks and 6 lights. On 3 July he got 29 (25 blacks, 3 lights); 23 blacks and 2 lights on 4 July; and 16 black moths (no light ones) on 5 July. The proportion of recaptured *carbonaria* was growing higher, as well as the total recaptures.

At the end of eleven days, 123 *carbonaria* and 18 typicals had been recaptured. The *percentage* of the marked releases recaptured – the presumed survivors – was what mattered: 27.5 per cent of the *carbonaria* versus 13 per cent of the typicals. Though perhaps not as dramatic as Hillary's feat, this was, in its way, a ground-breaking moment. Bernard had reportedly proven that black moths survived twice as well as light-coloured moths on the darkened tree trunks of a polluted wood near Birmingham. Thus they had a 50 per cent selective advantage. This unequal ability of individuals to survive and reproduce *was* natural selection. It would lead ineluctably to a gradual change in a population, with favourable traits piling up like compound interest over generations. Evolution, right before one's eyes.

As far as one can tell, no one has ever asked what happened between the last day of June and the first day of July 1953 to turn the tide. Table 5 reveals that after 30 June he began to release more moths. The average number released prior to 30 June was 30.8, while the average for 30 June–4 July was 95.2. Was this why he recaptured more? Or was there some other reason as well?

Michael Majerus thinks that the decisive factor may have been the weather – especially wind. 'I've had releases with several hundred moths and if the wind comes up, you never see one again – or you get the odd one. Kettlewell's moths may have just moved two miles to the east. It may well be the second of July saw a warm, calm period, a high pressure sitting over the wood, which meant that, first, lots of moths were flying, and, secondly, that nothing was dispersing very fast.'

To find out what weather conditions had actually been, I attempted to time-travel back to June 1953 via microfilms of *The Times* of London and the Manchester *Guardian*. As a vanished world reassembled itself and scrolled past the small viewer-window at breakneck speeds – Mau Mau outbreaks in Kenya, scary Cold War summits, soul-searching about 'commercial TV', photographs of the new Queen's triumphal state entry into Edinburgh in a coach pulled by four grey horses – I slowed down occasionally to savour the columns by Edmund Hillary ('How the Oxygen Problem was Solved',

'Alarming Gusts'). Eventually I became adept enough at the controls to bring the sometimes alarmingly speedy and clanky machine to a clean stop right on the page with the weather reports, which in *The Times* were tucked next to 'Today's Arrangements', the list of the royal family's activities that day. The weather in the British Isles between 25 June and 5 July, I learned, had been largely fair and rather warm, apart from a few scattered thunderstorms, some morning fogs, mists, and 'dullness'; winds had been mild to moderate most days. The *Guardian*'s reports told a similar story. As far as I could discern, the weather before 1 July was much like the weather after 1 July, if anything slightly better for moth collectors.

But weather was likely to be a local matter, and what I needed was Birmingham weather. Here the UK Meteorological Office came to the rescue by unearthing records for the Birmingham (Elmdon) airport 'for the hours of 0000z, 0600z and 1800z each day', plus the 24-hour maximum and minimum temperature, sunshine, and rainfall amounts for 25 June to 5 July 1953.[57] Since Bernard was suddenly getting more recaptures on 1 July and on subsequent days, one would expect to find the shift in weather, if there was one, on the 30th, when the first of those moths were released. The records show that the entire interval was dry and sunny, more like Phoenix than England. The barometric pressure was high and fairly constant during the whole period. As for wind, the calmest day was 26 June, the windiest was 29 June, but 30 June and 1 July were not appreciably calmer (and those days were gustier than 27 and 28 June). The winds did indeed die down a bit on 2 and 3 July, until the evening. July the 4th and 5th were about as windy as 30 June and 1 July had been. It is possible that we don't know enough about the very local weather conditions that might have affected Bernard's moths, but if the wind had been gusty for the first six days and subsequently died down, Bernard never mentioned it in any of his papers, or in his letters written from the field. It is the sort of thing he might have mentioned.

In the aviary Bernard had fiddled with his experiment, laying out, after two unpromising starts, a smorgasbord of bugs to get the results he wanted. It would have been natural for him to try to fix what wasn't working, just as he would have tinkered with a sputtering generator or changed the position of a cage in which two moths refused to mate. Is it possible that he made modifications in his experimental design in Beacon Wood? Working solo, Bernard made all the decisions about how many moths to place on each tree trunk, where to place the *carbonaria* and typicals, how far apart to space the moths. He alone rated the crypsis of each moth against its background and identified his catches by phenotype.

Another mystery concerns who was with Bernard in the woods. He uses the pronoun 'we' in his papers, but this may have just been a scientific convention, because otherwise, in his letter and reminiscences, there is no trace of anyone else – except for Hazel. She appears to have joined her embattled husband for a day or two, and on 1 July, the same day Bernard's recapture fortunes changed, his paper records that 'H.M. Kettlewell' – that is, Hazel – was gazing through field glasses at a tree trunk when something very important happened. She 'observed a bird fly up out of the bracken, snatch a betularia and return to the ground, the whole thing being over in a flash.'[58] No one had yet seen a bird in the wild eating a peppered moth off a tree trunk. Without this observation, any number of things could have been happening to the moths, including being eaten by something else. Thus this hedge sparrow was heaven-sent, and Bernard reported that he and Hazel were able to 'watch this bird regularly at work and to score the order in which it took its phenotypes'.

To bear out the industrial melanism hypothesis, birds had

to eat the conspicuous moths first – that is, the typicals. The first moth eaten by the hedge sparrow, on 1 July, according to Table 8, was a typical, with a score of -3 (very conspicuous). So was the second. The third moth was a *carbonaria* with a score of +3 (very cryptic); the fourth was a typical (with a score of -1). Two well-camouflaged *carbonaria* survived at day's end, Kettlewell reported. Later a robin was the predator *du jour*, alighting on twigs and bracken near the moth-festooned tree, 'whence it viewed the trunks and branches, making occasional excursions to pick up betularia'. After scooping up its prey on the wing, Bernard reported, the robin would fly to the ground to eat it, leaving wings and remains behind. It also ate typicals first and ignored, in every case, two *carbonaria*. As a result of their camouflage, these melanic moths would live to see another day, to mate and launch their genes into the future. They were the fit, the survivors, of this Darwinian drama. The results of Kettlewell's 'observation release' for seven days, that is, the census of moths remaining on the tree trunks at day's end when he made his rounds, were as follows: on average, 62.57 per cent of the *carbonaria* survived, compared with 45.79 per cent of typicals.[59] That, too, was just as one would have wished.

When Bernard Kettlewell packed up his car and trailer and drove away from Beacon Wood on 5 July – inadvertently taking away the Cadburys' keys in his pocket – he had reason to feel satisfied. He had found the predator of the original industrial melanic. The mechanism of microevolution was exposed, as if in a marvellous diorama: pale moths, dark moths, birds (the agent of selection) eating the moths, and all of it quantified. He had the numbers, and he knew they were good.

Of this spectacular success the world at large knew nothing yet. Only a handful of people knew what Bernard Kettlewell had done – Hazel, E.B. Ford, Cockayne, the Cadburys, a few collector buddies. Ford pronounced himself highly pleased: 'I am delighted with the results you have obtained during the past season. It is especially valuable to have established the predators of *betularia*.'[60]

As he set sail for Cape Town that September, Bernard was in Henry's good graces. His success reflected favourably on Ford's department, and in the annual report Henry delivered to the Nuffield Foundation that year, the peppered moth experiment was clearly the *pièce de résistance*. That autumn Henry was putting the finishing touches to his 'little book on moths'. He had written the chapter on melanism the previous February, but now that Bernard had 'cleared up one of the mysteries of that subject', Henry wrote to him: 'Will you allow me to scrap the passage [previously written] . . . and give a brief account of your own results in its place?' He added: 'The sooner you get this information into print, the more satisfactory it will be.'[61] In the normal scheme of things the world would have learned of the Birmingham experiment in a peer-reviewed article by H.B.D. Kettlewell, but here was his boss about to steal his thunder. Bernard could not refuse. Here is what Henry wrote:

Curiously, enough, it is an almost unrecorded event for a bird to pick a moth off a tree trunk, and, consequently, many entomologists have doubted its occurrence, save in quite exceptional circumstances. Yet were such selective elimination, *and on a large scale* [my italics] not a reality, the explanation of industrial melanism here advanced would be untenable. Indeed, the fact that it takes place *with great frequency* [my italics] was only established by the work of H.B.D. Kettlewell in 1953, and it then became clear why the destruction of moths in this way had not previously been detected.

. . . [Kettlewell] liberated 171 normal Peppered moths, *betularia*, and 416 of the black form, *carbonaria*, in the Cadbury Bird Reserve, which is affected by smoke-pollution from Birmingham. At that time, the titmice, *Paridae*, normally perhaps the most serious predators, were otherwise engaged, feeding on the tree-tops on the moth, *Tortrix viridana*. Having noted the positions of the Peppered Moths which he had released upon the tree-trunks, Kettlewell watched them from a distance with glasses and saw them destroyed in large numbers by Robins, *Erithacus rubecula*, and Hedge-Sparrows, *Prunella modularis*. The act of taking each insect was accomplished with extraordinary rapidity, almost in a flash, and could never have been detected without large-scale work of this kind.[62]

The vague and inflated terms he employed here – 'on a large scale', 'with great frequency', and 'destroyed in large numbers' – would soon get Bernard into some hot water.

Bernard should have been ebullient after his success, and his first letters from Cape Town that autumn were full of high spirits. 'Tell me – have we got another Lab boy?' he wrote to Philip Sheppard. 'How are Henry's circumstances? ... Did I tell you in my latest letter that I have found a larva on the mountain tops here which is so sensitive to sound that it jumps and shakes the plant and eventually throws itself off when a male voice shouts from a distance of four or five yards away ... When I made baboon-like noises it worked every time ... but never for Hazel.' Bernard wrote up his jumping and quaking caterpillar for the *Entomologist's Record*, speculating on the evolutionary reasons for this behaviour, but his discontent was growing. Once he had tasted the ambrosia of Oxford and felt the first frissons of immortalizing fame, Cape Town seemed a dull backwater, and by midwinter he was dejected and acutely missing his Oxford colleagues. A disorganized person by temperament, he was thrown into a funk by news of the departure of a certain Mrs Percy, an able lab assistant about whom Henry had written: 'she leaves us with some regret only because that miserable offspring of hers has reached some stage of its development that it needs lunch at home (I go without lunch) . . .' New uncertainties surfaced about Bernard's Oxford appointment and his tax status, but they were swiftly solved and by April a full-time appointment as Senior Research Officer at Oxford was assured. Now all the Kettlewells could settle in England permanently.

Bernard had planned to accomplish the second half (the unpolluted half) of his experiment in the summer of 1954, and had been breeding squadrons of moths for that purpose, but by autumn he was having presentiments that 'all was not well within the sleeve', that his pupae were full of virus, a doomsday scenario. When he returned to England in the summer of 1954, the weather was unseasonably cold, affecting all the insect life in unpleasant ways. Viruses were indeed rampaging through the *betularia* stock, killing everything. After the holocaust, he was obliged to ask collector friends to send him eggs to build up fresh stock free of virus. He was forced to delay his next mark–release experiment and set his sights on breeding enough moths for the summer of 1955.

The Thames Valley, where the dreaming spires of Oxford sit picturesquely between two rivers, may be one of the dampest pieces of real estate in England; bone-chilling dampness pervades everything. In the 1950s the ever-present low fogs mingled with the polluted clouds of Lord Nuffield's motor works in nearby Cowley to create a pestilential vapour that produced a high rate of bronchial ailments and depression. Bernard was constitutionally prone to depression anyway, and gravitated to warm, tropical places. For all its beauty, Oxford can seem a cold mistress, a *belle dame sans merci*. This Arcady is only for the elect, and if you don't belong, the thick medieval walls of the colleges, the portcullised gates, the cloistered gardens and the stone façades, turreted, castellated, and prickly with pinnacles, seem calculated to keep you out. Bernard Kettlewell was an outsider, an amateur without an advanced degree. His position was not a regular academic one, he did not have a college appointment, as most academics did, and he supervised no doctoral students. 'He had a real phobia about speaking about his science in public and would avoid formal teaching at all costs,' recalls his one-time assistant David Lees. Ultimately, he was there as a guest of E.B. Ford.

The chill seeped into Bernard's bones as he shivered in his caravan in Wytham Wood that July and August, putting up some twelve thousand *betularia* caterpillars in sleeves, hoping to 'get some through the season'. By September, apparently despairing of the peppered moth, Bernard was breeding the more sedentary industrial melanic moth *Gonodontis bidentata* in a frenzy, confiding to a friend that 'in view of the catastrophic effects of virus disease on *betularia*, it is absolutely essential that I repeat my pollution experiments as soon as possible on *bidentata* . . .'[63]

Their tax situation settled, the Kettlewells could stop living in trailers, and in the autumn they found a furnished rental house in Long Wittenham, near Oxford, to which Bernard always referred as 'ye olde Tudor cottage', and in which 'I bashed my head both fore and aft, got wedged in the bath [and] slept on a bed as hard as a board . . .' Around this time his spirits were restored somewhat by a major collecting coup in Romney Marsh, in Kent, where he got to show up a large party of entomologists who had been hunting in vain for a particular holy grail: the 'life cycle' of a moth called *Hydracea hucherardi*. This meant discovering the insect's foodplant (on which it lays its eggs) as well as larvae, pupae and adult moths. After the other lepidopterists left in defeat, Bernard identified the foodplant — marsh mallow — as well as the pupa and the perfect insect, and took some fertile eggs from a female to overwinter in glass-topped tins, from which he got larvae the following spring. When he published his discovery that January in *Entomologist's Gazette*,[64] he was hailed a conquering hero.

Bernard had also been on a quest for the right spot for his 'unpolluted' experiment. On a warm Sunday in July in rural Dorset, he stopped his car in front of an impressive set of gates leading to a vast, wooded country estate called Dean End Wood and obtained permission to use it for the second chapter of his historic experiment.

7

A most influential home movie

Word of Bernard's accomplishments had been circulating quietly among evolutionary biologists in Britain, and by early 1955 his name was becoming known within this small but elite world. The Royal Society had contacted him to put on an exhibition in connection with one of its so-called *Conversazione*, at which the rich donors and benefactors were entertained by interim reports from scientists at work. This was a considerable honour. Still, he had not published a word.

Unlike his boss, E.B. Ford, whose silken words streamed effortlessly from podiums or into dictaphones, Bernard found academic papers hard going. He was more at ease in the free-wheeling, off-the-cuff style of amateur publications such as the *Entomologist's Record*, where his article describing his successful quest for *H. hucherardi* had appeared in January. It took him more than a year to cobble together a draft of a serious scientific article about his Birmingham experiment, which then had to be vetted by everyone – E.B. Ford, Philip Sheppard, and even Sir Ronald Fisher. ('E.B.F. insisted on rewriting the introduction and summary and seems to have added a lot of "Ford 1940–41" etc. which I had inadvertently left out,' he confided to Philip. Henry wanted his own contribution to industrial melanism lavishly acknowledged.)

As it happened, Bernard's momentous achievement was first aired in an unusual and unfortunate venue. In the 15 March issue of the *Entomologist's Record*,[65] one of the editors, P.B.M. Allan, the well-known lepidoptery writer who had prescribed 'Brown Barbados sugar' as the *ne plus ultra* of sugaring, reviewed *Moths* by E.B. Ford and did not care for the Oxford don's patronizing tone toward 'the mere collector'. The bulk of his review focused on one chapter, 'Melanism, Industrial and Otherwise', largely devoted to Ford's interpretation of Kettlewell's still unpublished 1953 experiment. After reproducing the key passage about birds eating the moths 'in large numbers', Allan went on to quote Ford thus: 'Though much work of a similar kind remains to be done, the mystery surrounding the selective elimination of normal and melanic Lepidoptera *has been solved.*' Allan's italics, like arched eyebrows, telegraphed to his readers that he wasn't convinced.

Allan seriously doubted that a few birds could possibly eat so many moths, and he derided Ford and Kettlewell's explanation that no one had ever observed birds eating moths from tree trunks because they acted so quickly, 'in a flash'. He also pointed out that the fact that some moths failed to return to the traps didn't necessarily mean they were eaten by birds. He concluded sardonically: '. . . not only are the observations of this gifted observer [i.e. Kettlewell] always of value but on this occasion he has given a new orientation to our conceptions of the behaviour of both the robin and the hedge-sparrow.'

Bernard received his copy of *Entomologist's Record*, volume 67, on 15 March. He had just flipped it open and observed with pleasure that his urgent request for help with his *betularia* frequency survey had been duly inserted in the 'Exchanges and Wants' section, when his eye fell upon the 'Review of E.B. Ford's *Moths*'. His apoplexy can easily be imagined. Not only was he running into his first (severe) critic even before his journal article was published, but here was an old collector chum publicly doubting his word. 'Now it is one thing to challenge the interpretation of results,' he wrote indignantly to Allan that very day, 'but it is an entirely different one to query the records and integrity of the observer . . . Why on earth, if you had these doubts and misgivings, knowing me as you do, did you not write and ask me for fuller details instead of attempting to "shoot me down" in public . . . ?'[66]

He had been misquoted, Bernard said; he had never claimed that he 'saw them [the moths] destroyed in large numbers'. As for Allan's doubts about the birds' diet: 'A few years ago I would have agreed with you about the number of insects that birds eat, but I can assure you that, like me, you have a very wrong idea of their capacity . . .' If Allan had any doubts, he added, he could write to Niko Tinbergen (1907–88), the world-famous Oxford animal behaviourist. In fact, Tinbergen had just agreed to film his predation experiment the following July. He enclosed a copy of his forthcoming article in *Heredity*. Having read it, Allan agreed to print an excerpt from it in the May issue of the *Record*. The wording in *Moths* had been unfortunate, Allan said, and that was the fault of Dr Ford (from whom he had just received a long, hectoring letter lecturing him on the duties of a reviewer). Bernard was not completely mollified. Letters arriving daily from friends aghast at the 'sarcasm', 'implied dishonesty', and 'dangerous suggestions' contained in the *Moths* review only inflamed him all over again. It enraged him further when Allan wrote him: 'By the way, I think you ought to be very, very grateful to me (and I am sure you *really* are) for giving you the opportunity to dispel a misconception about your powers of observation . . .'[67] To this Bernard retorted: '. . . must point out that this has never been necessary until you thought fit to publish your unfortunate communication.'

As the *Moths* affair continued to simmer in the *Entomologist's Record*, Bernard was beset with other worries. Hazel's mother had become gravely ill, and Hazel went to Surrey to care for her, leaving Bernard alone in Ye Olde Tudor Cottage. At the same time, following a particularly vitriolic exchange with his old entomological mentor and partner, E.A. Cockayne, Bernard decided to cut off all contact with the man who had been something of a foster father to him. For an unloved, tyrannized only child, it was an extremely traumatic parting, and Bernard was very torn.[68] The Kettlewells were about to move, having found a handsome old stone vicarage to rent in Steeple Barton, just north of Oxford on the Banbury Road, on the edge of the Cotswolds. Not only did Bernard have to gather live peppered moths and other props for an upcoming Royal Society exhibit; above all, there were the Herculean labours of preparing for his pair of summer experiments, one in unpolluted Dorset and a repeat of the 'polluted' experiment near Birmingham.

Bernard already had too much on his plate and was taking it out on his children, secretaries, assistants, and anyone else in his frenetic orbit. It was his custom when feeling anxious to seek relief in the 'private lives' of wild moths and butterflies, so when his old friend Cyril Clarke asked for a little entomological help, he did not refuse. An eminent physician who had become enamoured of Lepidoptera, Clarke was doing serious genetic research on swallowtail butterflies (*Papilio*) at the University of Liverpool, where Philip Sheppard would join him in a year's time. The various *Papilio* polymorphisms were a crucial research subject, which would lead to key insights about the rhesus factor in human blood, among other things, and Cyril Clarke (later Sir Cyril) became famous for developing the 'Liverpool jab', an injection to save the lives of rhesus babies. This work was an outgrowth of E.B. Ford's early intuition that the human blood groups were a genetic polymorphism of great medical interest. To tease out the genetics of *Papilio* it was necessary to perform hybridization experiments, coupling closely related species and keeping records of the progeny, and Clarke wanted someone to travel to Corsica to collect *Papilio hospiton*, a rare swallowtail species he wished to mate with his *Papilio machaon*. Bernard simply couldn't resist the challenge.

Phobic about airplanes, he drove his Plymouth at a dizzying clip through the French countryside and took a ferry to Corsica, arriving to find the rugged mountains iced by a recent snowfall. The next day dawned hot and brilliant, and the tall, long-legged, middle-aged Englishman in shorts and knee socks must have been a strange sight striding through fields with a dozen assembling cages dangling from his belt. Failure was likely. His search of the literature had shown that no one knew much about the life history, time of emergence or 'localities' of these butterflies. 'I am still haunted by the awful thought,' he wrote Cyril, 'for some reason or other, should I fail to get *hospiton*, the disgrace would be too awful, and all I can offer is to do my utmost.'[69] Failure to catch a butterfly was, for Bernard, akin to a samurai's defeat. Fortunately, he soon came upon hundreds of gaudy *hospiton* males and females. They flew up into the central Corsican mountains in the morning, covering all the rocky outcrops and cairns at midday, then, as cool mists wreathed the mountains in the late afternoon, glided back down into the valley. Every day Bernard would chase them up the steep mountainside in the morning and back down into the valley in the afternoon.

He found the niches where *hospiton* lived, mapped out its life-cycle, identified its foodplants, caught live butterflies and hand-mated them to *machaon*. From the village post office he mailed live Corsican *Papilio* butterflies in damp packing to Cyril, explaining to the baffled postmaster that there would be more 'papillons' the next day. In Liverpool Cyril successfully bred them to butterflies from his broods and kept Bernard abreast of each stage of the hybrid offspring, from eggs, to larvae, to pupae, until, on 1 July, the first hybrid butterfly emerged, 'a most beautiful insect and strikingly intermediate'. Like gods, the scientists had fashioned creatures never seen on Earth before.

Bernard returned to England in early May a shadow of his former self, having lost so much weight dashing up and down Corsican mountains that even Hazel barely recognized him. Bernard was never more himself than in the field; and those who had seen him at work never forgot it. 'He was the best I ever met,' according to R.J. (Sam) Berry,[70] who worked as his assistant for several summers. Indoors, in a scholarly setting, he seemed a deflated, insecure version of himself, and this was more the case with every passing year. At Oxford his *betularia*, under the watchful eye of the laboratory assistant, seemed to be doing well as their hatching season approached, but Bernard was feeling strangely bereft. Some of the graduate students around the lab appeared to snicker at his 'countryman' garb and loud, jocular manners, and one priggish young man even slandered him to his own daughter, Dawn, saying that he did no work. Bernard knew he was not an intellectual and in no way equipped for the typical Oxford 'shootdowns' and verbal jousts whose rules sometimes harked back to the sixteenth century, as archaic as they were pointless. In contrast, Henry Ford, devastatingly articulate and oozing high culture, was made for Oxford. It was Bernard's secret belief, sometimes confided to a secretary, that the departmental eggheads with their 'firsts' were genetically unsuitable in many ways and quite useless in the field. Take the young Welshman, twitchy and 'brilliant', who slathered himself with insect repellant before going collecting in Wytham Wood and congratulated himself on not getting bitten by mosquitoes. 'I didn't catch any butterflies,' he had commented, mystified.[71] At times Bernard felt that he was standing outside a great restaurant, gazing wistfully through steam-misted windows at well-dressed people feasting on marvellous delicacies inside.

Oppressed by the fogs and miasmas of Oxford, he missed the sunny meadows of Corsica, the lure of a rare species, the hunt for aromatic foodplants. He missed joking with Philip, who was in the States, working on *Drosophila* in Dobzhansky's laboratory at Columbia. Henry was preoccupied with preparations for a trip to the big Symposium on Quantitative Biology at Cold Spring Harbor on Long Island, where important issues of population genetics would be aired. As usual, he would want to see which way the wind was blowing in the States and try to win converts to his viewpoint. He was deep in thought these days, doing his wounded partridge walk and muttering to himself, and had been overheard saying things over the phone like 'Excuse me, I didn't hear what you were saying because I was talking.'

'Selection experiments on industrial melanism in the Lepidoptera', by H.B.D. Kettlewell, came out in the prestigious journal *Heredity* in May 1955, but it felt a bit anticlimactic after the

Entomologist's Record affair. Bernard was also preoccupied. Not only was he busy wrestling with hordes of pupae for the summer's double experiment but he had to pull out all the stops to ready his exhibit for the big Royal Society soirée. This black-tie affair took place in London soon after his return from Corsica, and the Lord Mayor of London was among the important personages in evening dress who came to marvel at its central display, hundreds of live peppered moths, light and dark, arrayed on lichened and dark boughs. For some reason Bernard and Hazel left the premises briefly around 10 o'clock or 10.30, leaving Henry to watch over his exhibit. What happened next would become part of Bernard's legend. 'When I returned about 11 p.m.,' he wrote to Cyril Clarke, 'we had the extraordinary spectacle of seeing Dr E.B. Ford (whom I left in charge) pointing at the chandeliers whilst entirely surrounded by V.I.P.'s, explaining that Dr Kettlewell's exhibit had, unfortunately, transferred itself to the various lights . . .'[72] Whether this happened because Henry had neglected his duties, or whether it was just that Bernard's moths, being nocturnal, flocked toward the chandeliered light when it got completely dark, is unclear, but Bernard thought it was a hoot, and never tired of telling the story.

The family had just moved in to Steeple Barton and were still sitting on packing crates when Bernard set off for Dorset, one night in mid-June, to do his 'unpolluted' experiment. He liked to drive at night when there was no traffic on the road, and on this muggy night he was gliding through undulating countryside reduced to shadow-shapes under an overcast sky. By the time he reached Hampshire, there were flashes on the horizon, then a steady drumming of raindrops, finally loud claps of thunder. At three in the morning, the clutch burned out just as he was driving up a hill near Andover at the height of the storm. He tugged on his handbrake, but it could not hold both the heavy car and the caravan. He was forced to back both car and caravan back down the hill, a tricky manoeuvre that took hours and taxed his nerves, and there were no passing motorists to lend a hand.[73]

How he managed to get from Andover to Dean End Wood is not recorded. Perhaps the zoology department's Land-Rover was dispatched and hitched to the trailer with its incomparable cargo. We know from his letters that Bernard did not sleep for 48 hours, and that he did reach Dean End Wood, a lush woodland bordered on three sides by cornfields (inhospitable to moths and thus forming a natural barrier) and dripping with flowering creepers and velvety moss. 'It extends for miles,' he reported, 'is extremely pleasant and full of adders, deer, bird life, not to mention an enormous number of Roman Burial grounds.' Every tree trunk and branch was carpeted with greyish lichen, which the pale peppered moth's delicate wings seemed designed to imitate. Bernard had run some traps there one night in the autumn and caught 19 typicals and one *insularia*, not a single *carbonaria*. Leaf washings from local trees revealed minimal pollution.

It took hours to set up his tent and campsite, lug the heavy generator and mercury vapour trap to the centre of the experimental area, hang up all the assembling traps and unpack the sleeved pupae. What happened next is unclear. Perhaps the turmoil of the last twenty-four hours was too much, coming on top of the serial stresses of the preceding months – the move to Steeple Barton, the trek to Corsica, the overwork, his mother-in-law's illness, the Royal Society exhibition, the estrangement from 'Cocky'. Whatever the cause, Bernard's letters record that he suffered a 'most unpleasant turn' early in the experiment. It may have been a fainting spell, palpitations or irregular heartbeats, but it frightened him enough to speed off to the hospital in nearby Salisbury, where an EKG found nothing organically wrong with his heart, and Bernard was discharged back to his woods. His 'turn', he confided to a friend, 'was definitely the result of no sleep, stress and strain, and over smoking', and after it, he added, 'I had two others ["turns"] which put me back a bit, though fortunately I did not have to lie up, and so the experiments went continuously.'[74]

In Dorset in 1955 he followed essentially the same procedure as in Birmingham in 1953, except that he was now in unpolluted country where pale typicals should be favoured. He hoped to get mirror-image results, and he did. In his mark–release–recapture experiments, 13.74 per cent of the typical moths were recaptured, compared with 4.68 per cent of the *carbonaria*. That was three to one, even better than the two-to-one advantage of *carbonaria* in Birmingham. (His raw recapture figures here had actually worked out to a two-to-one advantage of typicals, but he decided, for reasons we'll discuss later, not to count several days, and this worked out to the figures above.) The two experiments, taken together, were a powerful demonstration of the value of crypsis. Light moths survived better in rural, unpolluted forests, melanics in polluted forests. Generations of students would learn that there was 'differential mortality', hence *fitness differences*, between the two forms of the peppered moth.

Having complementary experiments took care of a few problems. As Bernard was well aware, there were other possible explanations for his results. The lifespan of the two types, or 'morphs', might be different; alternatively, one morph might migrate farther and hence not return to the traps, or perhaps one morph was more attracted to light or to the scent of females in the assembling traps. But two experiments, one in which the light moths survived better and one in which black moths did, convinced him that selective predation was the sole explanation.

Yet something else was needed to clinch the case: the 'agent of selection'. Industrial melanism occurred, Bernard proposed, when well-camouflaged black moths acquired an advantage in industrial areas by being less conspicuous to *predators hunting by sight – that is, birds.* There was an entrenched belief among lepidopterists and ornithologists, however, that birds were not natural predators of peppered moths, and Bernard's somewhat anecdotal reports hadn't satisfied everyone. Until the agent of selection was nailed down, it was possible to suppose that other things were happening to the typicals that didn't make it back to the traps in Birmingham, and to the *carbonaria* in Dorset. Thus, still smarting from P.B.M. Allan's sarcasm and determined to prove his enemies wrong, Bernard had brought a secret weapon to the woods of Dorset. This was the Dutch-born Niko Tinbergen. Tinbergen, who had spent two years in a Nazi concentration camp for opposing the expulsion of Jews from the universities, was one of the founding fathers of ethology. He would share a Nobel Prize with Konrad Lorenz and Karl von Frisch in 1973; his brother Jan won a Nobel as a founder of econometrics in 1969, making the Tinbergens the only pair of Nobelist brothers. Tinbergen was considered a seer of animal behaviour, had his own school of ethology at Oxford, and was, moreover, a pioneer in the making of nature films. Knowing that he had recently made a film of birds, Kettlewell asked if he would be kind enough to 'cine' birds eating his moths. Wooed by Bernard's charm and intrigued by the project, Tinbergen readily accepted.

When Bernard set out his moths on tree trunks in the morning, he laid out an especially dense cluster at the proper height for Tinbergen's camera. After a breakfast of Bernard's fried kippers, Tinbergen repaired to a canvas hide with his camera, while Bernard followed his routine, checking on his pupae, monitoring the moths he had set out on trunks and tallying the missing, slogging through the masses of insects in his mercury vapour trap and counting the relevant ones. The birds behaved as Bernard had hoped. While under observation, spotted flycatchers, nuthatches, yellow hammers, robins and thrushes gobbled up peppered moths by the dozens − and they ate more of the *carbonaria*. During one session spotted flycatchers, observed by Tinbergen, consumed 46 black moths and only 8 typicals. In another period, Kettlewell watched them put away 35 *carbonaria* and one typical. The other birds feasted in a similar fashion on the lavish banquet and were caught in the act, both on film and in still photographs. (In his paper, Kettlewell would admit to putting out huge concentrations of moths on these closely watched trees in order to get the photographs.)

Camping under the stars with hundreds of moths from Oxford, Bernard and Niko forged a schoolboy camaraderie that the normally straitlaced Tinbergen would count as one of his finest adventures. Perhaps he had not camped with anyone as flamboyant as Bernard; maybe he was inebriated by his wild sense of humour, his silly jokes about the sex lives of moths, or the potency of the cocktails he made every evening. One day the two men collected wood raspberries and soaked them in a bowl of sherry and cider, stashed the fermenting concoction under the car, and forgot it for three days. When they finally drank their wild raspberry wine it was potent in the extreme. Still more intoxicating, perhaps, was the realization that they were the first people on Earth to catch evolution in action − on film!

What the biology textbooks would fail to mention, and what passed unnoticed by their peers for at least a decade, was that Bernard had done a little tweaking in this experiment, too, just as he had at the aviary and in Birmingham in 1953.

* * *

While Bernard was counting moths in Dorset, his boss was in fine fettle in America. Unlike Bernard, Henry adored travelling by 'flying machine', and for a man whose sensibilities were largely rooted in the late nineteenth century, if not in classical Greece, he was strangely mesmerized by the Manhattan skyline, pronouncing the new glass and steel towers of Rockefeller Center, without irony, 'the most wonderful thing that has been built in the world since the Middle Ages'. In general Henry enjoyed a sense of well-being, a balm for his nerves, in 'my beloved New York', as he did in his other favourite city, Amsterdam.

Science historians would later maintain that the Cold Spring Harbor Symposium XX of 1955 was population genetics' finest hour. Among the sparkling bays and lush woods of Long Island's north shore the kings of population genetics strode past magenta clouds of wild rhododendron, and gave talks that would reverberate for the next decade of evolutionary science. Henry entertained himself by taking delight in his opponents' shortcomings. His old adversary Sewall Wright, with whom he and Ronald Fisher had nearly come to blows over *Panaxia* a few years earlier, was more than usually opaque, and did not even offer the comfort of a clarifying graph, and Henry noted that he had shrewdly switched his terminology from 'random genetic drift' to 'accidents of sampling'.

Many of the participants would say that the symposium's final three sessions were the most rewarding. The first of these, on 'Variability and Polymorphism', was dominated by E.B. Ford, who laid out the conditions that he believed permitted especially rapid evolution to take place, citing his own studies of the variation of the meadow brown butterfly on the Scilly Islands. Following Ford to the podium, Anthony Allison of Oxford divulged new details of his research on the genetics of sickle cell anemia. The hemoglobin molecule in red blood cells exists in two forms. The mutant form, which differs from the normal form in a single amino acid substitution, results in sickle cell anemia in homozygotes, people born with two copies of the defective gene. However, in the heterozygous 'carriers', with just one copy of the gene, it causes few or no symptoms and confers protection against *falciparum* malaria, a deadly disease prevalent in West Africa. This finding was important to the cause of natural selection because it seemed to answer the age-old riddle of how mutations, which were largely harmful, could somehow provide the raw material for positive evolutionary change. Here was a highly deleterious, even lethal, gene that offered an advantage in the heterozygous form, thus maintaining a balanced polymorphism indefinitely. By the end of the decade sickle cell anemia and the peppered moth would be the two staples of every biology textbook, and E.B. Ford would cite Allison's work almost ritually.

The symposium was also noteworthy for the leading talk by Theodosius Dobzhansky, one of the fathers of the Synthesis, in which he defined the war between the 'classical' and 'balance' hypotheses. The classical school, embodied by Thomas Hunt Morgan and his caste of fruit-fly adepts, believed that for every gene there was a normal or 'wild type' and that all variations were aberrations. The classic-view geneticist H.J. Muller, a star graduate of Morgan's 'fly room' at Columbia, had, at a 1949 conference, floated the influential concept of 'genetic load'. Nearly all mutations were bad, he said, and would be removed by natural selection, while the rare favourable mutations would rapidly become the new normal 'wild type'. Most genetic variation, therefore, was transient. Humans, alas, had increased the rate of mutation and decreased the operation of selection through civilised improvements in public health and medicine, and the species was deteriorating under the insupportable 'genetic load' (too many harmful mutations in the gene pool). In his eugenicist tract 'Out of the Night', Muller had advocated, among other things, mass artificial insemination of women with the sperm of men superior in intellect and character.

Dobzhansky could not have agreed less. He'd put in his time in the field, catching fruit flies in the High Sierras, and, like most field workers, was convinced that the diverse and ever-changing character of natural environments required a great storehouse of genetic variability. Variation was indeed the norm. Selection generally acted not to sift out bad mutations but to preserve variation. The 'balance' school insisted there is no single 'wild-type' allele and that natural populations tend to be highly heterogeneous – an insight that has indeed been borne out for most populations.

His good friend E.B. Ford agreed. If Dobzhansky and Ford and their respective protégés became fixated on the concept of *heterosis*, or heterozygote advantage, it was for this reason: if heterozygotes are generally fitter than homozygotes, then genetic variability is *good* for individuals as well as species. Also, heterozygote advantage was the supposed outcome of Fisher's theory of the evolution of dominance. (Today it is generally conceded that, in Michael Majerus's words, 'there are relatively few known examples of heterozygote advantage, and the evidence that this is a widespread phenomenon is still lacking.')

When Henry returned to Oxford in mid-July of 1955, Bernard was up to his elbows in mark–release–recaptures at his old polluted spot, the Cadburys' estate, near Birmingham. He was repeating the 1953 experiment on a smaller scale, this time with Tinbergen filming the birds. The departmental secretary, Bridget, briefed Henry on everything that had happened in Dorset – the broken clutch, the good numbers, the 'turns' and chest pains. 'I do greatly congratulate you, and I can assure you that the results which you have obtained are causing widespread interest in the U.S.A.,' Ford wrote to Bernard upon his return to Oxford in mid-July. Bernard's 'turns' worried him, however, and in his best avuncular manner, he urged Bernard not to 'be a fool', to 'stop overdoing things', to rest for a few weeks.

'I absolutely insist, Bernard, that you do not stay up late at night working, but that you rest and have long nights in bed. If we can get the Dr Tinbergen over to you to take a cine film, we will do so . . .'[75] Bernard, however, was constitutionally incapable of moderation in the field. The June nights are short at latitudes of 50-plus, but even during the dark hours Bernard's brain was buzzing. In an effusive memoir,[76] Tinbergen recalled: 'Often he did not return from his last round until well after midnight. And even his few hours of sleep were restless. Time and again he would suddenly wake up and call out:

"Niko!"

"Bernard?"

"Can you hear the generator?"

"'Yes, of course. Listen."

"Oh yes. I thought it had stopped. So sorry."'

No wonder he steadily lost weight in the field.

After releasing marked moths for two days, and counting recaptures for four, Bernard had the singular good fortune to get ratios nearly identical to those of 1953. Of the *carbonaria*, 53.25 per cent were recaptured; of the typicals, 25 per cent. The percentage of returns was higher than in the previous experiments, and Bernard could not say why, but the black moths still had a two-to-one advantage, just as before. Then there was the film. For two days Tinbergen squatted in his hide filming birds – redstarts, in this case – picking moths off tree trunks. (The moths had been placed there 'in quantity', and replenished when all of one morph was devoured.) During these sessions, Tinbergen observed birds eating 43 typicals and 15 *carbonaria*, and, Bernard would report, 'on the majority of occasions, two or more typicals were eaten before a *carbonaria* was discovered.'[77] Tinbergen's camera was recording the proof that birds ate the moths and they ate the non-cryptic ones first.

Niko Tinbergen and his film camera were fated to silence all of Bernard's detractors for ever – or practically for ever. If the man who had so brilliantly deciphered the reproductive behaviour of gulls said birds ate peppered moths no one dared argue. (It should be noted that Tinbergen also believed that childhood autism was caused by 'refrigerator mothers', and in his 1983 book *Autistic Children: New Hope for a Cure* floated the theory that autistic children could be cured by a flurry of maternal hugging.) Besides, there were the photographs, and the film. The still photos came back from the lab in mid-July, and Tinbergen was pleased with the results. 'The nut-hatches are perhaps the best bird shots of all,' he reported, 'but of all five species I have what I need: robin, yellow hammer and songthrush first looking up at tree, then flying up, then snatching a black moth off, then down on the ground again with the moth . . .'[78]

There is a minor mystery around Tinbergen's role. In a little memoir, 'Happy Moments with Bernard Kettlewell', written in 1979, he recalled spending *three weeks* with Kettlewell, in *Dorset*. (He doesn't mention Birmingham at all.) Yet Bernard's letters from the field convey a sense of beleaguered solitude, and if Tinbergen was already there, why would Ford have written to Bernard: 'If we can get the Dr Tinbergen over to you to take a cine film, we will do so . . .'? It seems fair to surmise that Tinbergen actually accompanied Kettlewell in the field no more than a week in all, a few days in Dorset, a few in Birmingham.

Tinbergen's film had its premiere at an intimate little gathering at Miriam Rothschild's house. In the midst of the screening, Henry suddenly exclaimed in his high voice: 'I have just seen an animal come out of the bookshelf.' Tearing their eyes away from the blurry birds and lichened trunks, the audience fastened on the beady eyes of one of the Lane children's pet rodents, appearing just behind the screen. Close examination revealed that it had eaten a path through the length of the bookshelf, tunnelling through a multi-volumed set of *Lepidoptera of the World*, to emerge in mid-film. The flickering black and white images were not much more polished than a home movie of a child's birthday party, but the effects were galvanizing, as if Galileo had had a friend with a super-8 filming him dropping weights at Pisa. Here was evolution literally in action, captured on celluloid. The critics were instantly trumped, and Bernard, savouring this I-told-you-so moment, mischievously penned a caption for the film that read: 'This film is dedicated to P.B.M. Allan, O.B.E., whose doubts inspired it.'[79]

Only the surly Cockayne saw fit to grumble, in the terminal phase of their friendship: 'The Cadbury Reserve [meaning the 1955 experiment] is just amusing. It was all artificial and coloured pictures of a redstart are no more convincing than your [illegible] stories of robins and hedge sparrows.'[80] Cockayne evidently did not believe in the moth-gorging birds, and he was not inclined to be generous at this point, for Bernard had just severed their connection. ('It is a tragedy that I had to take this step,' Bernard confided to a friend, 'but . . . I could not take the continuous abuse I received in his letters . . . I must admit I find it very much less of a strain now that I no longer receive those hurtful letters.') Hazel tried to mediate, but when she went to Tring to talk to the sickly old man, he remained intransigent and rude, and, learning of Bernard's spells in Dorset, he insisted, coldly, that his former protégé was obviously 'slowly cracking up'. Bernard indignantly wrote that he was not, and the two men never spoke again before Cockayne's death in 1957. Even after his mentor was in the ground, Bernard was still complaining that Cockayne had kept and donated to charity several valuable prints belonging to Bernard.

After all the 'stress-strain', Henry ordered Bernard to 'do as little work as possible for the next two or three months'. The Kettlewells were still settling into the vicarage at bucolic Steeple Barton, in a shallow valley on the fringes of the Cotswolds; some ancient clause in the tenancy required Bernard to read the lesson in church on Sundays. Here he was the proverbial big fish in a small pond, a famous scientist in a tiny English village, and the lifestyle of a local squire suited Bernard. The vicarage was rather more house and land than the Kettlewells could afford, but they had got a bargain on the rent because it was unfurnished. There were several acres of land, and a trout stream next to which Bernard planted a garden of larval foodplants and set up his mercury vapour trap. There was a formal perennial garden and a vegetable garden with hand-lettered signs addressed to the avian life: 'No bird larger than *Parus major* allowed.' Soon, the bushes and tree boughs would sprout swelling muslin bags teeming with caterpillars. The garden would be home to a rather famous artificial colony of scarlet tiger moths (*Panaxia dominula*) and over the years young men from Oxford would appear at all hours to change sleeves, collect foodplants, or net rare mutant moths.

It is a curious fact that virtually all discussions of industrial melanism up to the present time cite E.B. Ford as the originator of the theory Kettlewell tested in the field. But Ford's original theory was rather different. In 1937 he had proposed that 'melanic forms have spread in industrial areas owing, primarily, to selection for characters other than colour'. By raising broods of caterpillars under conditions of semi-starvation, a standard technique, he found that the black forms were physiologically hardier, and he reasoned that they would have the selective advantage whenever their colour did not work against them. Thus crypsis would come in only as a secondary factor, for if colour alone determined the change in frequencies, why hadn't melanism occurred in all moth species in industrial areas? 'It is remarkable,' observed the formidable zoologist Arthur Cain, who while at Oxford had done the influential *Cepaea* experiments with Philip Sheppard, 'that even Ford had hardly as yet achieved the idea of strong visual selection *for* the melanic form in polluted areas. Kettlewell and Tinbergen's demonstration of strong selective predation must have come as a surprise to both Ford and Fisher. Kettlewell himself told me that he was assured, by lepidopterists and ornithologists alike, that birds did not take moths resting on tree trunks, certainly not in polluted districts.'

Bernard's new status in the scientific community was unveiled publicly in December at a Royal Society 'discussion' on 'The Dynamics of Natural Populations', which Henry had arranged to celebrate 'remarkable advances in the field' and a growing 'realization that, in certain conditions, evolution takes place in natural populations much more rapidly than had previously been suspected.' This was his repetitive theme, but now he had Bernard's moths as ammunition. Bernard's talk, 'A résumé of investigations on the evolution of melanism in the Lepidoptera', was clearly the main event, and he probably stammered and spoke too fast, as was his habit when nervous, before the roomful of FRS-bedecked worthies like David Lack, J.B.S. Haldane, C.D. Darlington, Kenneth Mather and Julian Huxley.

Huxley adored Bernard's work. He proclaimed to the gathering that, whereas Ford, Fisher and Sheppard, working with

Panaxia, had been 'unable to assign a biological reason for the alteration of advantage and disadvantage [in other words, did not discover what the fitness advantages consisted of] ... Kettlewell has shown how the melanic morph of the Peppered Moth ... enjoys a cryptic advantage in industrial areas, but a disadvantage in non-industrial ones.'

If there was a sting for Ford in finding himself upstaged he kept it to himself. He would manage soon enough to put Bernard in his place.

8

A reptile's leg into a bird's wing

Bernard had carried out his experiments at just the right time, for evolutionary biology seemed to be on the cusp of solving all the great questions. By the close of the 1950s, the peppered moth would be *the* poster child for evolution – 'Darwin's missing evidence' – and Sir Gavin De Beer, director of the British Museum, would write that 'evolution is nothing but a byproduct of improvement of adaptation. Biologists are . . . fortunate in that they have a cast-iron example of how adaptation has arisen, under man's eyes in the last hundred years. It is the Peppered Moth *Biston betularia* . . .'

Evolution had been redefined as a shift in gene frequencies during the Synthesis years, and now natural selection was being worshipped as an almost omnipotent force. Much as natural theologians of the nineteenth century had sought signs of a watchmaker God in the 'intricate adaptation of form and function to environment', as Stephen Jay Gould would wryly point out,[81] so now the evolutionary geneticists scanned the 'intricate adaptations' of organisms for evidence of the all-powerful hand of natural selection, which came to seem like a god. Julian Huxley asked rhetorically, in 1953: 'It seems to be true that natural selection can turn moths black in industrial areas, can keep protective coloration up to a mark, can produce resistant strains of bacteria and insect pests. But what about elaborate improvements? Can it turn a reptile's leg into a bird's wing, or turn a monkey into a man? . . . In a word, are you not asking us to believe too much?'[82]

The answer, according to Huxley, was no. The epic progression of life from unicellular organisms to humans required no new laws, no abrupt leaps or transubstantiations: it was built of the same basic stuff as the observable genetic shifts of peppered moths. Macroevolution – evolution above the species level – was simply microevolution writ large; the apparent qualitative differences among species and higher taxonomic groups were the inevitable culmination of quantitative changes extended over time, though it might require a brief period of geographic isolation for varieties to diverge into different species. Thus large-scale events like the evolution of major groups of organisms, such as the cat family, were no more than an amplification of processes such as mutation and natural selection that could be observed in the field or laboratory. The story of the peppered moth was evolution in miniature. The 1973 biology textbook *Life on Earth* by Edward O. Wilson *et al.* asks the student to ponder the following:

Did birds really arise from reptiles by an accumulation of gene substitutions of the kind illustrated by the raspberry eye-colour gene [in *Drosophila*]?

147

OF MOTHS AND MEN

The answer is that it is entirely plausible, and no one has come up with a better explanation consistent with the known biological facts. One must keep in mind the enormous difference in timescale between the observed cases of micro-evolution. Under natural conditions the nearly complete substitution of the melanic gene of the peppered moth took fifty years. Evolution of the magnitude of the origin of birds usually, perhaps invariably, takes many millions of years ... The reading from [the fossil record] ... leads to the conclusion that [evolution] is based upon hundreds of thousands of gene substitutions no different in kind from the ones examined in our case histories.[83]

Although there had been other experiments demonstrating natural selection, most of these were flawed or missing something vital. 'Getting an experiment that showed natural selection in action, and one that was a very good example, was very hard to do,' explains Cornell biologist William Provine, the pre-eminent chronicler of twentieth-century evolutionary biology.

If you think of all the other examples of natural selection between, say, 1890 and the 1960s, the truth is that there isn't much. You have Bumpus's experiments on sparrows. He doesn't know dingle about the genetics of sparrows, so that doesn't have any great persuasive power. You've got Weldon and his crabs – Gosh, I have a letter from Arthur Cain agreeing that it was a very flawed example.

In the Forties we got Fisher and Ford's work on *Panaxia*. That was a wonderful experiment and Sewall Wright said it showed absolutely nothing. When Cain and Sheppard did their work on *Cepaea*, it was considered to be one of the best examples of natural selection. Then Camille Lamotte did the same stuff in France and didn't get their results at all. Cain and all these people came up with what they called 'area effects', which took away from what they did earlier. It wasn't clear at all. Allison's work on sickle cell anemia seemed so clear and obvious, but it turned out to be so much more complicated than it looked on the surface. It was a hornet's nest; it was awful.

So Kettlewell's experiments had the makings of what everyone had been hoping for – the example of natural selection in action. And it completely captured all the textbooks. It's fun to look through all the textbooks and always this example – and I mean *always* – is hauled out.[84]

In the winter of 1955, years before the textbooks began to display his triumph, Bernard shut himself up at Steeple Barton with his new dictaphone to dictate his second paper for *Heredity*, the one that reported his results from Dorset. His dictating style drove every secretary mad, for after speaking into the machine for some time, he would blurt: 'No, cross that out and start again . . .' oblivious to the fact that she would have already typed a page on bulky layers of aubergine-coloured carbon.

The School of Ecological Genetics had a single departmental secretary, whose job it was to answer the phone, type everyone's correspondence, scientific papers and book chapters, as well as feed hungry larvae their proper foodplants. She was quartered in an outbuilding, a collapsible garden shed known as 'the hut', which was even smaller, shabbier and more remote than the 'insect house' where larvae and pupae ate and slumbered in their pots and tubs. Whenever the phone rang, she would have to walk outside, cross a gravel path – slushy in wet weather, wearing a thin crust of snow in winter – to enter the insect house, where the telephone was.

'Are you good at getting along with eccentric people?' the personnel office would ask candidates in their interviews. They had to be. Serving two masters, tight, fastidious Henry and helter-skelter Bernard, the job had a split personality. Henry would dictate letters or book chapters in perfect, sinuous compound sentences; Bernard's thoughts would gush out in a muddle. Whereas Henry's papers were neatly filed, all of Bernard's reprints, first drafts of chapters, bank statements and personal letters commingled in a single fan file. 'Bernard and Hazel couldn't put two bank statements together,' remembers Ruth Wickett, who worked as departmental secretary in the 1970s. 'I had to help them.'

Shuttling back and forth between the two bosses required the tact of a courtier, as each man considered his work the highest priority. Though his discomfort with 'female women', as he continued to refer to them, was palpable, Henry was generally courteous and thoughtful, always writing a note to the Hope Department librarian, Audrey Smith, before he visited the library. 'Nobody else did this; they would just walk in, but Dr Ford would write me from his club in London, "Dear Miss Audrey . . ."' Smith remembers.[85] His strange little games could also make life difficult for the staff. 'If you rang him up on the telephone, he'd answer the door, and if you knocked on the door he'd answer the telephone,' recalls Kate Coldham Davies, who was the departmental secretary from 1958 to 1962. 'It was his way of not communicating with people. One day I positioned myself outside his door and told someone to ring him. So the telephone rang and he answered the door and I was standing there. After that I had no trouble at all.'

Bernard tended toward heedlessness and frantic disorganization. It was typical of him to dictate long letters over the telephone from Steeple Baron, requiring his secretary to stand for an hour or more in the insect house, cradling the phone awkwardly in the crook of her neck while taking painful dictation. Since he often worked at home, he would blow into the offices unpredictably like a sudden gust of wind, often with a funny story to tell. 'One day he saw a woman drop her pocketbook in the road. All the contents had spilled out, and he helped her pick them up, and some of them were quite embarrassing,' Kate Davies remembers. 'He went on and on about it.' Another day, roaring in from Steeple Barton, he stopped at a garage and asked the attendant to refuel his caravan. 'What caravan?' the attendant asked. It had swung loose along the way. Whenever he arrived, he would instantly commandeer all the secretarial help. If he could not find his secretary – if she had ducked into the lavatory – he would bellow for her at the top of his lungs from the foot of the stairs.

'He was a very big man. Everything about him was big, his voice, his gestures. He wore open-necked shirts and these awful navy blue serge shorts, like scout shorts, and knee-length navy socks and sandals. Even when he wore a suit he managed to look slightly rumpled; there was always cigar ash on his lapels. He chain-smoked Wills Whiffs, small cigarillos. There were clouds of cigar smoke, and everything he wore smelled of it. He used to use the nicotine from his cigars to kill the moths – he'd put a little bit on the end of a pin and put it through the moth.

'He had these huge hands but he could do such delicate things, like hand-pairing these butterflies. Imagine hand-pairing butterflies! You can do it, but you have to know what you're doing.'

Bernard's work required him to constantly tinker, improvise and invent – light traps, bottled pheromones, radioactive tracers, special sleeves, made from laundry bags, for confining caterpillars in their peat- and privet-filled flower pots with the precise amount of moisture. The file cards recording his breeding records were handled as if they contained the secrets of the philosophers' stone, and daily messages between Bernard and his technicians give a hint of how single-minded and painstaking life in the laboratory was: 'Humuli – changed food, larvae fine. Peltigera – 4 moths hatched, transferred to marigold plant. Looking each day for sign of eating. Betularia in hutment are not very good as you know. Most have got disease.'

His voice was so loud, especially on the telephone, when he seemed to be bellowing all the way to America, that members of the department used to measure sound volume in 'Bernards' and 'mega-Bernards'. He'd often bound into the lab with dogs in tow; Henry, who detested dogs, would overlook their presence as long as they were described as 'pussies of the canine variety'. (Henry was apt to speak of all animals thus, as 'pussies of the ovine variety', 'pussies of the bovine variety', and so on.) Where Henry savoured contrived donnish jokes, Bernard was a schoolboyish prankster, who once, in the course of a special exhibition at the museum, filled a glass case with a foul stinkhorn fungus and a bottle of Airwick, wrote a notice about it in rhyme, and put it on display.

Everyone in the department – especially the secretary who had to function as his muse – knew that translating his ideas into scientific prose was exquisite agony for Bernard. The 1956 *Heredity* article, describing the second set of experiments, done in 1955, had not been too awful an ordeal, perhaps because Henry and Philip had stepped in as midwives, but the write-up of his survey '*Biston betularia* in Britain, 1952–1957' paralysed him for the better part of a year. The editor of *Heredity* demanded several rewrites, and Bernard complained bitterly about having to express himself in 'telegram English', until at last 'A survey of the frequencies of *Biston betularia* (Lep.) and its melanic forms in Great Britain' was published in 1957.[86]

This was a crucial part of the industrial melanism story,

for it was vital to show that differences in fitness between the two morphs of the peppered moth were, in fact, driving evolution. If crypsis and differential predation by birds were really responsible for the rise of industrial melanism, there had to be a correlation between the *fitness* of either morph in a given region and its *frequency*. If melanics were twice as 'fit' in his predation experiment in Birmingham – that is, they survived twice as well as typicals – then melanics should also outnumber typicals to a corresponding degree in samples of the wild population. Given this correlation, it should ultimately be possible to predict the frequencies of typicals and *carbonaria* in different geographic regions, knowing the level of pollution, the extent of destruction of lichens, and so on. (Bernard's work would thus meet one of the criteria of a good scientific theory – namely, that it should generate predictions that are capable of being falsified.)

For years Bernard had been engaged in the vast enterprise of sampling the different morphs of *Biston betularia* all over Britain. He had unearthed old data from museums and private collections, private records and entomologists' notebooks, and since 1952 had personally been dashing around with his traps to sample dozens of different areas. This had the added advantage of getting him out into the field, where he was happiest. He couldn't cover the whole country singlehandedly, of course, so he assembled a network of eighty amateur entomologists (their ranks eventually grew to 150) who agreed to run light traps in their gardens every night and send him the *Biston* data from their districts. Thus he had frequencies of melanics from the Isle of Man, Scotland, Cornwall, East Anglia, Essex, Cheshire, Yorkshire, Lancashire, Cumberland – just about everywhere.

Perhaps only Bernard could have carried this off. His mystical bond with the amateur community, the fact that he spoke its language and knew its habits and secret handshakes, was something E.B. Ford could not have duplicated in a hundred years. The amateurs, in turn, would not have minded that their curious and solitary hobby could acquire transcendent significance through the Oxford man's work. His connection with them was complex and consuming, encompassing exchanges of 'material' (eggs, larvae, moths) and breeding data, as well as trap counts. A 1955 letter to a collector, F.W. Jeffery, of Plymouth, captures the flavour:

Unfortunately, the small male moths which would have been most useful for me were dead on arrival . . . I think that the inside of your box is too smooth, so I am getting my Secretary to send you half a dozen pill boxes enclosed inside a tin, which you can send them in, as we know they travel well this way. I shall also be sending you a *carbonaria* female . . . and I am wondering if you could be so good as to assemble a male to her, and possibly you or some of your friends would like to breed from her, thereby guaranteeing that this generation is free from our Laboratory virus, and it would be of the greatest interest for us to see if next year any of the *carbonaria* showed a tendency toward an insect intermediate between *carbonaria* and typical, though it may take several more generations of out crossing to obtain this.[87]

His 1957 article in *Proceedings of the Royal Society* contained an extensive chart of melanic frequencies in 83 centres in Britain, which seemed to support the general thesis of industrial melanism. High frequencies of *carbonaria* were found near industrial centres such as Manchester, Birmingham, Sheffield and Liverpool, and low frequencies in rural districts such as Cornwall, Devon and rural Wales. Western Britain was vitually melanic-free; every one of Bernard's informants in Cornwall caught 100 per cent typicals in his traps. Northern Scotland had no melanics but the Glasgow district had 90 per cent. In Ireland there were few melanics apart from a handful around Dublin and Belfast. All this was just as expected, but, puzzlingly, a high frequency of melanics turned up in *rural* eastern England, an anomaly Bernard tried to explain as 'the indirect effect of long continued smoke fall-out carried by the prevailing south-westerly winds from central England'.

Reviewing the history of industrial melanism, and drawing on Haldane's and Fisher's calculations, Bernard identified three stages in the rise of melanism.

(1) During the first period, which he called the 'adjustment' period, the melanic mutation would have appeared and then spread quite slowly through the population until it reached the level of 1 per cent. This period would have taken about 38 years, he calculated. Only at around the 1 per cent level would the mutant come to anyone's attention, he thought. Assuming that had happened in 1848 in Manchester, the original mutation would have occurred in 1810.

(2) During the next period, of 'rapid rise', assuming the selective pressure is maintained, the melanic gene would spread very rapidly until 50 per cent or more of the population consisted of *carbonaria*. Under certain conditions the frequency would rise from 10 to 70 per cent *carbonaria* in a mere ten to fifteen years.

(3) The next period would be one of 'slow elimination of f. [form] *typica*' or, alternately, 'a balanced polymorphism'. Bernard still hadn't made up his mind which way things would go. So far he could report that 'in every industrial area f. *carbonaria* is now at least 85 per cent, but in no large sample is the value 100 per cent. Even after one hundred years, Manchester and Sheffield will have 1–2 per cent of non-*carbonaria* forms . . .'

The whole thing was represented by a lovely sigmoid curve, first suggested by Haldane in 1924. One central mystery about industrial melanism was its origin. Were all the dark peppered moths in Britain the descendants of a single mutant near Manchester? Or did separate, recurrent mutations in different places keep supplying *carbonaria* genes? Bernard waffled. '. . . it is indeed difficult to resist the conclusion that the new mutant has radiated outwards from an original centre of mutation near Manchester,' he wrote. 'There is no doubt that following the original capture in 1848, the nearest counties to Lancashire were the next to record its appearance; Cheshire in 1860, Yorkshire in 1861, Staffordshire in 1878, and Westmorland in 1870.' Two paragraphs later, he asserted that 'dispersal cannot be accepted as the actual cause of the widespread distribution of f. *carbonaria*. There is considerable evidence that recurrent mutation, at a fairly high rate, also takes place. *Carbonaria* had constantly appeared in isolated centres separated from others by usually impassable barriers.' Few people at the time seemed to mind Bernard's having his cake and eating it, too.

This was, without a doubt, the period of Bernard's 'rapid rise'. His film – or rather Tinbergen's – had been seen and admired by sixteen hundred people at a recent Royal Society event; the BBC had been calling; and Bernard had been asked to put together a permanent exhibit for the British Museum. He was invited to join the elite Entomological Club, which was so exclusive it had only six members. He had entered a stage of life when discussions of film rights, royalties, contracts and television broadcasts would consume the bulk of his correspondence. His face began to appear on television, as a staple of certain nature films, sometimes alongside the popular British nature-show personality Armand Dennis, whom he resembled uncannily. (During his African period Bernard had been all alone one day, he thought, admiring the view from a remote peak in the Belgian Congo, when he felt a soft female arm creep around his neck. 'Darlink, isn't it vunderful?' sighed a silky voice which was revealed to belong to Michaela Dennis, Armand's wife, who had mistaken Bernard for her husband.[88]) As the peppered moth became a household word, this unselfconsciously racist ditty made the rounds at Oxford:

> There once was a fellow named Kettlewell
> Who found out why moths become black as Hell
> When soot is ejected
> Black genes are selected
> 'Cause birdies can't see darkie moths as well.[89]

To the world at large, he was the famous Dr Kettlewell of Oxford, whose work was eulogized, both in his lectures and writing, by the great E.B. Ford. It would have astonished most people to find how much more uncertain Bernard's stature was within the sub-department of Ecological Genetics itself, for by this time it had become plain to those around him that statistics and genetics were over his head, that he didn't quite think like a scientist, and that he was prone to rather elementary gaffes when he tried to discuss evolutionary theory. Oxford life continually demanded of Bernard a set of skills he lacked.

Some people found him fun, lovable and charismatic; others were not so charmed. The plaque on his door read 'Dr H.B.D. Kettlewell', but of course he was the wrong kind of doctor for Oxford, and one morning some wag had scrawled the message 'He's not really a doctor.' 'Kettlewell was a bit of a joke,' says Laurence Cook, who was there from 1957 to 1960. 'He didn't teach, so he wasn't part of that setup. He was a specialized person; a lot of people in the department never even met him. He was this bluff, blustery chap who used to rag on people probably because he felt very shy and ill at ease in the department.' Many people in the department were aghast at the way he berated his lab assistants and secretaries. J. James (Jim) Murray,[90] who worked next door to Bernard, recalls: 'He was always swearing at his secretary, telling her she was no good, screaming "Why did you do that?" I didn't like him.' Once, finding himself in Henry Ford's office as the roar of one of Bernard's tirades wafted through the wall, David Jones remembers Henry saying calmly: 'Bernard is a man who generates emergencies.' Those who were fond of him said his bark was worse than his bite. 'He could be a marvellously kind and warm host, full of humour and bonhomie,' recalls his sometime assistant David Lees. 'Equally, he could be a bully and make a lot of noise to get this own way. I discovered that if one resisted his more overbearing characteristics, he quickly climbed down, like all bullies.'

At the beginning of their alliance, Ford assumed the role of benefactor, solicitous about Hazel and the children, helpful about tax matters and references for landladies, urging 'dear Bernard' not to overwork, to take care of his health. He was fulsome in his praise, courtly in his thank-you notes. As time went on, Ford the perfectionist began to find more and more faults in the man he had hired and, while always praising him to the world beyond Oxford, within the ivory tower he treated him as something of a buffoon. 'It was,' according to David Kettlewell, 'a love-hate relationship. When my father came back from South Africa to Oxford, it was because Henry needed him. My father always felt beholden to him for bringing him to Oxford and accepted his idiosyncrasies.' Sensitive to slights and always desperately insecure, Bernard became increasingly intimidated by Henry's basilisk gaze and his nuanced but lacerating put-downs. The only time he was seen without a cigarillo in his mouth was when he ventured into Henry's office, where he would chew peppermints nervously. When he emerged he usually looked deflated, as Kate Davies observed. 'The way Bernard spoke about Henry, the look on his face after he'd been in to see him, made me feel it was a very difficult relationship.'

David Kettlewell echoes that account: 'Henry would come by and say, "Bernard, I want to see you." My father would go talk to Henry and come back looking fully dejected. Henry had a way of being abrupt; he'd talk down to my father in front of other people, though there was nothing overtly offensive about it.'

'Ford was a terrible bully!' as Miriam Rothschild knew well. 'I once saw him bully someone so badly I almost got up and left the table. He bullied Bernard, who had all these insecurities. Once a situation is set up where one person bullies another, the bully is just as stuck as the one being bullied.' In this case, both Henry and Bernard were stuck with each other.

As an amateur, Bernard did not always master the idiom of science, and he frequently betrayed himself in his papers where others would have known how to cover their tracks. In writing up his aviary experiment he had failed to conceal the false starts he made before hitting on the experimental design that worked. And the first paragraph of the 1955 *Heredity* paper contained the following interesting admission: 'Yet after more than twenty-five years of observation and constant enquiry, I have found no single instance in this country in which anyone has witnessed a bird detecting and eating a moth belonging to a protectively coloured (or "cryptic") species when sitting motionless on its correct background.' It would be several decades before anyone in the scientific community thought more about what that sentence meant.

At this point only old John W. Heslop Harrison, who had originally traced industrial melanism to a mutation induced by industrial smoke, spoke out strongly against the new evolutionary talisman. In a churlish and somewhat incoherent piece in the *Entomologist's Record* in 1956, he disputed Ford's claim (articulated in *Moths*) that the mystery around industrial melanism had been 'solved' by Kettlewell's experiments, and insisted that 'the real problem, the cause of the development of melanism in the insect concerned, has never been touched.' Kettlewell/Ford traced the phenomenon to a random mutation that they hypothesized had popped up periodically in the past and disappeared again, but around 1850, when the industrial revolution created favourable conditions, gained a foothold and spread through the population by means of natural selection. Heslop Harrison disagreed strongly, insisting that 'there must be an inciting agent responsible for the appearance of black mutants in the first place . . . I believe . . . that melanism has been induced by a melanogen or melanogens present in industrial smoke, and that such inductions are recurrent.'

He argued with many of Kettlewell's methods and results, and claimed that he himself had placed *Biston typica* and *carbonaria* side by side on lichen-covered bark, with results opposite to Kettlewell's: 'In every trial made, the typical form was, by far, the more conspicuous.' He had even graver doubts that birds normally ate resting moths; for years he had kept track of melanic and typical forms of the moth *Antitype chi* on walls near his home in North Durham, and always deemed it 'an extraordinary happening if a single specimen of either form disappeared' during the day. And why hadn't Kettlewell looked into predation by *bats*? 'In [Ford's] view, they take no part in bringing about the preferential elimination of types because they feed on the wing. I think that position is completely indefensible. Both in the woods and elsewhere bats do take considerable numbers of various species of moths. Surely, in the twilight of a darkening wood the lighter individuals would be more conspicuous, and therefore the more likely to be captured? This indubitably would be Natural Selection in action. Thus, if birds are to be invoked as responsible agents in a winnowing action, bats cannot be neglected.'[91]

In 'Melanism and an answer to J.W. Heslop Harrison', in the next issue of the *Entomologist's Record*,[92] Bernard conceded that in his own experiments 'I was plagued with bats to such a degree that I had to take precautions against them. Male *betularia* were eaten in scores as they flew to the assembling cages, yet there was never any evidence that they took more of the lighter individuals than of the dark which, had it been so, would, of course, have been reflected in my recapture figures.' Then he moved on to ridicule 'the Professor's' embarrassing neo-Lamarckian views. 'The indisputable fact is that he believes that environment (food or chemicals) affects the "soma" which in its turn changes the "germplasm", and that this is then inherited according to Mendel's Laws.'

Of course, everyone understood that this idea violated the most sacred law of genetics: the separation of phenotype and genotype, of 'soma' and 'germ-plasm'. Around the turn of the century August Weismann had formulated what became known as the Weismann doctrine, that a zygote is the starting point for two independent processes, one leading to the 'soma', the individual organism (which can be affected by external conditions), the other to the inviolate 'germ line', the sex cells that are the seeds of the next generation. The latter is a one-way street. Acquired characteristics cannot be inherited. No matter how much the giraffe might stretch its neck, its chromosomes remain untouched by the experience and its descendants will not have longer necks. Francis Crick would state this 'central dogma' in molecular terms, explaining that information can flow from nucleic acids (in the chromosome) to proteins, but not the other way around. Bernard's friends congratulated him on having won the debate. He had the force of the *Zeitgeist* on his side. John Heslop Harrison, FRS, had been an important scientist with a long string of publications and awards, but by this time his induction experiments were suspected of being fraudulent or, at the very least, dead wrong, and he would go on to discredit his name with a series of bizarre botanical frauds on the island of Rhum, where he evidently planted imported species that he would later 'discover' as native.[93] 'Heslop Harrison has not the first glimmerings of a scientist,' Ford sputtered to Bernard, 'and now he is in his dotage.'[94] It was easy to discount his arguments.

Since the early to mid-1950s the marvels of evolution had begun to figure in the gee-whiz scientific/technical optimism of the era, alongside the miracle kitchens displayed at World's Fairs, the cute whizzing electrons of the Atomic Energy Commission logo, the first stirrings of space travel. Evolutionary themes seeped into popular culture, and technological progress and evolution became linked themes in science fiction. 'Arthur C. Clarke's *Childhood's End*, 1953, drew heavily on the work of [the Cheshire science fiction writer] Olaf Stapledon, who in turn had been inspired by [H.G.] Wells, who had collaborated with Julian Huxley on the *Science of Life*,' notes science historian Betty Smocovitis.[95] Everything was getting better and better all the time.

Darwin had been unable to convince himself that the mechanism of natural selection had any progress or purpose built into it. 'The basic theory of natural selection offers no statement about general progress, and supplies no mechanism whereby overall advance might be expected,' Stephen Jay Gould points out, correctly, noting that this 'denial of progress' was an extremely 'outré' idea in the nineteenth century.[96] Even well into the twentieth, the Synthesis leaders were inclined to speculate loftily about the 'future of Mankind', and the emerging evolutionary humanism, espoused most vociferously by Julian Huxley (who became a radio celebrity on 'The Brains Trust' programme), held that human beings could and should consciously build a better future as agents of their own evolution. Theodosius Dobzhansky, who, unlike Huxley, was not an atheist, mused in 1955 that natural selection 'leads to increased harmony between living systems and the conditions of their existence'. On another occasion, he lectured: 'It is Man the Geological Agent, not Man the Species, who will be having the greatest influence on the Earth. Man has now reached the point where he is adapting his environment to his genes rather than permitting his genes to adapt to his environment.'[97]

This belief in evolutionary progress was one reason that natural selection had to be very powerful and that nondeterministic processes such as random drift had to be rejected. If it was all just tosses of the dice, where could progress come in? The faith that selection was all-powerful everywhere was what Stephen Jay Gould would come to characterize as 'the hardening of the Synthesis'. He meant that as neo-Darwinism became dogmatic, no forces or laws other than Fisher's relentless engine of natural selection were allowed to matter much. This calcification took place wherever there were evolutionary biologists, but particularly in Darwin's native land. 'England has produced a long line of hyperselectionists,' Gould would note, 'from Wallace in Darwin's own day, through the purity of R.A. Fisher's adaptationism, to the convictions of E.B. Ford and the Cepaea school launched by A.J. Cain.'[98]

Bernard Kettlewell was no philosopher, and Henry Ford certainly lacked the epic scope of a Huxley or a Dobzhansky, but their blue-ribbon exhibit – industrial melanism – became an avatar of all these ideas and aspirations. One frustrating feature of evolutionary biology is that selection is so difficult to observe that 'nothing short of enormous efforts and considerable resources would suffice to detect it in any one particular case,' writes Michael Rose in *Darwin's Spectre*. 'And this is more or less what Darwin expected.' Therefore, evolutionary biologists came to rely heavily on inference. According to evolutionary theory, the average 'fitness' (or average rate of increase) of a population tends to increase when selection is occurring. Therefore, according to Rose: 'If we find the fitness of a population increasing from generation to generation and do not know anything else about the organism, the natural conclusion would be that selection is occurring. If at the same time, we notice the frequency of a particular attribute, such as darker colouration, increasing at the same time, it would be reasonable to assume that darker colouration, or some attribute related to it, was being selected for. We have thereby indirectly inferred the action of selection.'

A physicist would mop his brow at all these assumptions and inferences, which generated a widespread research style of looking for a feature (a certain colour or a certain mating behaviour, say) that *looks* adaptive and then trying to pin down the mechanisms of selection that might have generated it. The reasoning is *post hoc*, starting with an end result and reasoning in any way possible to trace it back to a predetermined cause. As we have seen, the selective mechanisms in the case of E.B. Ford's scarlet tiger moths remained obscure, and the snails weren't all that much more illuminating. Industrial melanism, in everyone's mind, seemed to be a cut above all the other examples from nature, for the mechanism of selection was known and quantified – or so everyone believed.

'It used to be argued that natural selection was only conjecture, because it had not actually been witnessed,' a 1966 book raved. 'By now we have become aware of many examples of natural selection in action. One of the best documented cases deals with the peppered moth.'[99]

The peppered moth was becoming evolution's number one icon just in time for the big Darwin centennial. The year 1958 was the hundredth anniversary of the reading of the Darwin/Wallace papers before the Linnean Society; 1959, of the publication of *On the Origin of Species*. Among other festivities, *Life* magazine sponsored Bernard Kettlewell to travel to Brazil 'in the footsteps of Darwin' in the spring of 1958 and photograph 'what Darwin might have seen'. (En route he and Hazel had made a stop in the Canary Islands, where the noisy revelry in their room one night brought a portly German tourist clad in a silk dressing gown to their door, complaining he couldn't sleep. 'You hideous Kraut!' Bernard exclaimed to the astonished man. '*You* kept *me* awake for four years, from 1941 to 1945.') In Brazil the jungle heat melted his flash apparatus, but Bernard managed to take still photographs and movies of 'extreme insect adaptations', some of which appeared in a lavishly illustrated *Life* spread entitled 'Darwin's insects – jewels of his jungle paradise'. Bernard gagged at the frothy text, which had Charles Darwin arriving 'wide-eyed, straight from the misty downs of Kent' to contemplate 'the mysterious laws of nature', and complained to a friend: 'I have repeatedly pointed out to them that the whole object of the exercise was to show that the laws of natural selection are not a ruddy mystery. Why must these Americans write stuff like that?' More arresting than the insects was a photo of Bernard stripped to the waist and covered with moths. When it turned up next to an article about his work in *Time* magazine a year later, he was mortified.

The principal centennial celebration should properly have occurred on English soil, but the University of Chicago became the venue after it managed to lure Sir Julian Huxley, the grandson of Darwin's 'bulldog', with an endowed position, sabbatical leave, and first-class airfare for himself and his wife.

It was a symbolic coup for the American Society for the Study of Evolution, rather like getting London Bridge, and, as the icing on the cake, Darwin's mathematician grandson Charles came too. This all but guaranteed that the rest of the British evolutionary establishment would follow, except for Ronald Fisher and J.B.S. Haldane, who were not invited. Nothing could better testify to the changed fortunes of Darwinism than the contrast between the subdued 1909 anniversary in England and this supercharged extravaganza, which encompassed five days of pageantry, televised debate, Darwin worship and theatrical spectacle. The manic atmosphere was such that one speaker was struck dumb and had a 'religious experience' that required him to be hospitalized for three weeks.[100]

Henry pronounced the Windy City a 'bottomless pit of a town, vast, black, sordid, and cheerless', but he was animated on his 'panel', enthroned alongside Ernst Mayr, Theodosius Dobzhansky, G. Ledyard Stebbins, Sewall Wright, and other heroes of the Synthesis. They all chatted amiably about microevolution, natural selection and polymorphisms, and nobody seemed to argue very much, perhaps because by this time they all agreed on the big picture. Natural selection ruled. It was as if the Oxford School of Ecological Genetics had taken over the world's brain. Henry cited industrial melanism as a situation *par excellence* 'in which natural selection is operating rather powerfully'. The adaptationists were still wrestling with the problem that all the novelty of evolution had to arise from mutations, which were usually injurious, and Huxley could point to the peppered moths as 'a very good example of how a mutation that is bad in some circumstances may be good in others. The small mutations that made black moths blacker would originally have been deleterious, but, once there was a black form under positive selective pressure, they were advantageous.'[101] Bernard didn't go to Chicago, but his and

Tinbergen's film *Evolution in Progress* impressed a large and receptive international audience.

On Thanksgiving day, Huxley, as the convocation speaker, declaimed: 'Evolutionary man can no longer take refuge from his loneliness in the arms of a divinized father-figure whom he has himself created, nor escape from the responsibility of making decisions by sheltering under the umbrella of Divine Authority . . .' He added that 'evolutionary truth' would free humans from 'subservient fears' and show the way to 'our destiny and our duty'. It was quite like Sir Julian to go on in this way, in flowery secular-humanist rhetoric, but in the American Midwest his 'secular sermon' (delivered in a chapel!) did not go down particularly well, especially with the press, and Huxley's atheism and the general Darwinist pep rally were noted darkly by a small group of outraged evangelicals. A stream of anti-evolution literature followed, notably John C. Whitcomb and Henry Morris's *The Genesis Flood*, the forerunner of the 'scientific creationism' movement, which in another three decades would pounce hungrily on the flaws in Kettlewell's experiments.[102]

It was fitting that the full mythologization of the peppered moth should have coincided with the centennial year, for Kettlewell's experiment brought evolution to life in a fashion so simple even a schoolchild could grasp it. When the article 'Darwin's missing evidence' by H.B.D. Kettlewell appeared in *Scientific American* in the same year, the peppered moths came to the attention of a popular audience in America for the first time. The catch-phrases coined in this article – 'evolution in action' and 'Darwin's missing evidence' – would inevitably be repeated verbatim in hundreds of other articles, and in textbooks. 'Had Darwin observed industrial melanism he would have seen evolution occurring not in thousands of years but in thousands of days – well within his lifetime,' the article concluded. 'He would have witnessed the consummation and confirmation of his life's work.'

Bernard stressed the same themes in 1962 when he showed his film at the tercentenary celebration of the Royal Society in London to an audience that included the Queen Mother, in a tiara and a full-skirted evening dress. He had savoured being photographed earnestly explaining industrial melanism to the Duke of Edinburgh at a previous reception, and had been preparing for this second royal moment for two years, enlarging photographs of moths, arranging title cards, dragging in logs, lichened and blackened, from the countryside. He had been denied permission to collect moths inside the royal gardens, but by collecting outside the gates he had been able to capture a genuine Buckingham Palace melanic, which was now entombed with its pale counterpart inside a special glass paperweight with a suitable inscription for the Queen Mother. Whether or not she had ever heard of industrial melanism, the Queen Mum fixed her brightest regal smile upon Bernard as he lectured. Afterwards, when Hazel and Bernard were presented to her, she had something nice to say and no doubt left feeling good about having seen something very much like evolution in a bottle.

Yet Bernard's growing fame had not really made life any easier for him at the only club he really cared about – Oxford. It wasn't hard to figure out that his rages at his staff reflected his frustrations and that his sometimes juvenile clownishness masked fears and insecurities that ran deep. Like many a comedic personality, he was in fact a man of sorrows, and the chief of these was his daughter, Dawn, the family's problem child. When the 1956 Hungarian uprising was brutally suppressed by Soviet tanks, Bernard wrote to the Red Cross offering his services as a physician to help Hungarian refugees. 'At the same time,' he wrote, 'I would like to hear from you whether you could find any uses for my daughter age 19 who is just completing her course in shorthand and typing. She is efficient at cooking, and reasonably so at organizing, and it seemed to me that she might come out with me at the same time.' Transparently, his real purpose was to find a niche for his problem daughter, who had been drifting since leaving school in 1954.

In the end neither Bernard nor Dawn worked with Hungarian refugees, and Dawn could be found intermittently giving demonstrations of cookery around Oxford. Beautiful, wilful, and sexually promiscuous, she had given her parents many a sleepless night since her early adolescence, and at Oxford her name was soon found associated with lewd graffiti on university notice boards. In those days Dawn's affliction went by the name 'nymphomania' (today it would probably be considered a subtype of obsessive-compulsive disorder), and various psychiatrists were consulted for it, but no one seemed to know what to do. 'She was insatiable,' says someone who knew her. 'Bernard was always tearing his hair out and it was terribly embarrassing at Oxford because the stories were always getting back to him.'

Young David, meanwhile, was sent to Malvern, a prestigious public school. Bernard had always dreamed that his only son, whose head he had filled with natural history lore since babyhood, would take up where he left off and become a scientist at Oxford. The pressure on David must have been intense as it became increasingly clear that he was not cut out for scholarship. He was an attractive and engaging boy, but his marks were disappointing, and tutors were always being engaged, in some desperation. When it was time for him to leave Malvern in 1962, Bernard briefly considered sending David to a 'crammer', a school designed to get a lagging student back on the Cambridge or Oxford track, before despairing of Oxford and sending him to a technical school in Cheltenham.

Bernard was thin-skinned to begin with, often detecting slights when none were intended (as well as when they were) and reacting badly to the highbrow teasing that was the favourite indoor sport at Oxford. Of all the scientists in the Oxford orbit he most admired Philip Sheppard; once when Henry forgot to mention that Philip was visiting from Liverpool, 'Bernard became hysterical and looked as if he were going to cry,' Henry reported to Philip. During another of Philip's visits Bernard joined a party for lunch at an Oxford restaurant and found himself impaled by Philip's wit. 'After chiding me in your own particular manner,' he complained to him the next day, 'I attempted to enter the conversation on Mimicry. You immediately implied that I was only qualified to speak about Industrial Melanism . . . I suggested that no-one particularly wanted to talk about Industrial Melanism, and it was this remark which lead [sic] to mine about you working on it . . .'[103]

Sheppard's sharp tongue was a familiar part of the landscape, which his colleagues took in their stride. 'You could become a Fellow of the Royal Society,' Austin P. Platt recalls Sheppard telling him during his year at the University of Liverpool. 'All you have to do is win a Nobel Prize.' But Bernard's insecurities made him less resilient than most. By this time it may have begun to dawn on him that he was a one-trick pony, and the charge that he had no standing to speak about other important evolutionary matters, such as mimicry, went right to the quick. He had 'solved' industrial melanism and propelled his name into all the books alongside the giants of evolutionary biology, but in the eyes of Oxford he wasn't one of them and never would be. ('It's a club,' says Laurence Cook, 'and the ways of chucking someone out can be very subtle.')

The luncheon brought up another sticky subject: Sheppard and Cyril Clarke were trying to replicate Bernard's experiments on the peppered moth. 'Something you said,' Philip wrote to Bernard, 'suggested that you resented Cyril and myself working on industrial melanism. I wouldn't have taken your remark seriously except for an incident two years ago when you told Cyril and myself that if we published a certain paper on our result you would never speak to us again.' He hadn't resented their working on the subject, Bernard retorted. 'What I did resent was the "shoot-down" impression you gave in the first paper you showed me . . . You may remember me saying . . . "What on earth would people think when a co-worker and friend like yourself write in this style."'[104]

In the winter of 1963, as Steeple Barton, snowbound for a month, fell under an arctic spell, Bernard had too much time to brood. He was bogged down in a biography of Darwin that Julian Huxley had talked him into co-authoring. Huxley, who tended to flit between his multiple interests, had become distracted in the meantime and hadn't written a word, leaving Bernard holding the bag. What Bernard really should have been writing, Henry was fond of reminding him, was his *magnum opus* on industrial melanism.

Increasingly, Bernard was working at Steeple Barton instead of facing his demons of Oxford. 'At midday, but not before, he would normally have his first large gin and tonic,' recalls David Lees, who frequently found himself at Steeple Barton dealing with the sleeves. On one occasion when he was invited to lunch, Lees saw Bernard disappear into the downstairs lavatory, and an alarming explosion ensued. 'It turned out that Bernard had spotted some wood pigeons on his cabbage patch and the lavatory window was perfectly placed to shoot the marauding birds. Bernard kept his twelve-bore in there and it seems this was a regular occurrence. It explains both why excellent pigeon pie was on the menu at Steeple Barton and how it came to be that I found lead shot in my brussels sprout.'

In June Bernard tamed a young robin in his garden and made it his chief taste tester. 'He flies onto my hand and takes chosen insects offered by me,' he wrote to his friend Haldane. 'The red of *Arctia caja* (the garden tiger which Helen [Haldane's wife] once ate in public at the Lamb and Flag!) was sufficient to teach him to leave untouched Burnet moths, all other Tigers, and even the red-banded Burying Beetles . . . He gets furious if, when hungry, only a Garden Tiger is offered! – and he was an egg 8 weeks ago!'[105] Robins were not the only taste testers. Bernard thought nothing of eating the odd moth, nor did Miriam Rothschild nor (evidently) Haldane's wife, Helen. Kate Davies once came into the office to find Miriam Rothschild and Bernard 'hunched over a table eating Burnet moths. "Want one?" they asked. "No, thanks," I said, "I had lunch."' While the palatability of insects to predators was an important evolutionary issue, public moth-eating was also a display of field machismo, part of the get-your-feet-muddy culture of entomology. The most ostentatious moth-eater of all was Henry Ford. Anyone who had seen him, with studied casualness, pick up a moth and pop it into his mouth, chewing meditatively as the powdery wings disappeared between his thin lips, would not soon forget the sight.

When Bernard published an article in the *Entomologist's Journal* summarizing his twelve-year survey of peppered moth frequencies in the British Isles,[106] Henry and Philip were appalled. Bernard had not bothered to update his tables and had simply reproduced the old, out-of-date figures, and they worried that his amateurishness might reflect badly on the School of Ecological Genetics. 'The whole thing,' Philip chided him, 'gives the impression that you couldn't care less about this particular piece of work and merely took the 1958 paper, put a

1. Charles Darwin believed that the operation of natural selection was too slow to witness in a human lifetime.

2. When the flamboyant J.B.S Haldane calculated the selective advantage of the melanic peppered moth in 1924, fellow scientists were stunned by the idea that evolutionary change could occur so dramatically.

3. Ronald Aymer Fisher, shown here with is young sons in 1926, determined to make Darwinism a rigorous science. His theories encouraged the young E.B. Ford to attempt to prove natural selection in the wild.

4. Manchester, the site of the first observed instance of industrial melanism in the mid-nineteenth century, was still very polluted in 1954.

5. and 6. This is the classic set of peppered moth photographs, taken under Kettlewell's direction at Oxford, that became a persuasive tableau of 'evolution in action'. The moths are alive but artificially posed.

7. The University Museum, Oxford, housed the laboratories of the Oxford School of Ecological Genetics.

8. The spires of Oxford, with All Soul's College in the foreground. As a fellow, E.B. Ford savoured its recondite traditions and opposed the admission of women.

9. E.B. Ford, the father of ecological genetics, was always nattily and formally attired. He took Bernard Kettlewell under his wing, but their relationship was often rocky.

10. Bernard Kettlewell collecting in the Belgian Congo, 1950.

11. The Kettlewell family at the beach in South Africa, 1950. They would come to regret moving back to England.

12. Kettlewell's famous moths attracted royal attention twice. His chat with Prince Philip at a reception in London in 1960 was a source of pride.

13. Julian Huxley's atheistic 'sermon', delivered from the pulpit of the University of Chicago's chapel, alienated some Americans, and inspired the first wave of 'scientific creationism'.

AXELROD   DOBZHANSKY   FORD   MAYR   EMERSON   HUXLEY   NICHOLSON   OLSON   PROSSER   STEBBINS   WRIGHT

14. When E.B. Ford, Julian Huxley, Theodosius Dobzhansky and other grey eminences discussed evolutionary theory at the 1959 Darwin Centennial at the University of Chicago, the peppered moths were invoked several times.

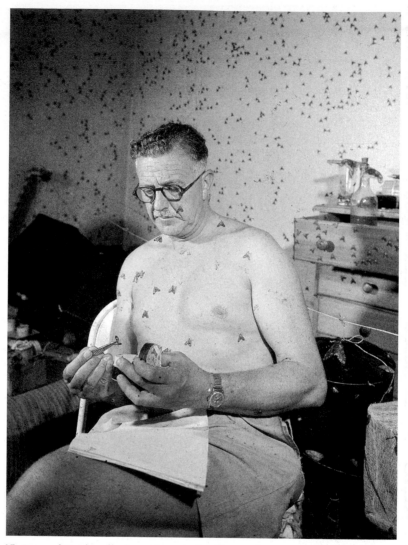

15. Bernard Kettlewell, shown here during his 1958 journey to Brazil, was never happier than when surrounded by exotic lepidoptera.

16. and 17. These photographs, or ones very similar, can be found in many current textbooks, purportedy illustrating the perfect crypsis of pale speckled peppered moths on light trunks, melanic moths on dark trunks. These moths are clearly dead, however; their spread wings show that they are pinned specimens posed on a tree trunk.

18. Ted Sargent. For years, he was the only scientist to see the flaws in the textbook case of 'evolution in action'.

19. The University of Massachusetts.

blue pencil through sections of it and republished the rest.' Not realizing that Henry and Philip had already had one of their periodic 'what to do about Bernard' talks, Bernard complained to Henry about Philip's 'dictatorial manner' and explained that his article was merely a 'trivial letter I wrote to pacify all those people who have helped me'.[107]

As Bernard's increasingly jagged relationship with Henry undermined his self-confidence, he felt alternately desperate to regain his boss's favour and resentful at his subservient role. There were days when he was as touchy as a jilted lover, blurting out his grievances, complaining that Henry had no time for him. 'I think Bernard is going the way of Arthur Cain [who was famously touchy] at his work,' Henry wrote Philip. 'He suddenly began speaking to me in the most extraordinary way (accusing me of not being friendly to him) quite suddenly, in the middle of an ordinary phone conversation the other day.'[108] Bernard was prickly, but it is a good bet that the phone conversation was not as innocent as Henry claimed.

Ford demanded total obedience, according to Miriam Rothschild. 'Everyone had to kowtow to him. Once he asked me if I would look at one of his papers. I thought he genuinely wanted me to make suggestions, but he just wanted to hear what a wonderful paper it was. And it was.' Bernard, who had been his own boss before coming to Oxford, was not confortable salaaming to Henry all the time, yet he was as dependent on his chief's patronage as any Renaissance artist or astronomer was ever at the mercy of a Medici prince. He had given up his medical career and a comfortable income, uprooted his family, risked everything, and now he was stuck in a deeply unhappy professional 'marriage' from which there was no possibility of divorce.

This atmosphere of unceasing intellectual warfare was one of several things that Hazel hated about Oxford. As a former

Birmingham Debutante of the Year, she was accustomed to lovely manners, yet at Oxford she was condemned to sit on the sidelines while young men argued incessantly and rudely about science. It was rare that anyone addressed any words to her. These conversations seemed to her to have no point beyond demolishing someone's arguments, scoring points, and demonstrating one's cleverness. The memory was no less painful for their son David: 'I would sit in the corner and listen. It seemed as if they were having terrible arguments all the time. Everything anyone said would provoke an argument about whether you were right or wrong. You had to have all your facts in front of you.'

When Julian Huxley or J.B.S. Haldane came to dinner it was different, for they were fun and truly liked Bernard and Hazel. When Henry Ford visited, in contrast, he struck the Kettlewell children as a peculiar, effete and rather cold figure, and they had to stifle their giggles when he ritually made a beeline for the kitchen, doffed his bowler hat to Nanny, and said: 'Good evening, Nanny. How is your pussy?'

Bernard would take him aside and say: 'Henry, you can't say *that* to her,' and Henry would say: 'I don't see why not. I am simply asking Nanny about her pussy.' It was one of the ways that Henry played the fool.[109]

Hazel felt equally alienated by the horsey set around Steeple Barton – people who looked down on one from horseback and whose conversation never strayed much beyond the subject of fetlocks, sires and gymkhanas. She had never wanted to leave Cape Town for this Siberia, but had done it for her husband's sake. Having believed she had married a doctor, she now found herself an entomologist's wife, which was a step down, socially and financially. 'She made an enormous sacrifice,' says David Kettlewell. 'Father always understood this, and was very grateful to her.'

Friends could see that it was Hazel who really called the shots, and that, without her, her husband's psyche would undoubtedly unravel. 'Hazel was charming, a strong person, had high standards and expected them of others,' says David Lees. 'She was one of the few women that Ford seemed to acknowledge, even respect. I think Hazel and Nanny ganged up on Bernard occasionally and together probably manipulated him very well.'

For Henry 1963 was a banner year. He completed two of his major goals: becoming a Professor of Ecological Genetics and bringing out *Ecological Genetics*, published by Chapman and Hall, the *magnum opus* he claimed to have conceived in 1927, when he had a sort of vision after reading an oracular paper by Fisher on mimicry. Henry's book was dedicated to Sir Ronald, who had died a year earlier. The aging Fisher had been even more helpless than usual when he had visited Henry for a week in 1961. 'I can manage to look after him and keep things reasonably right for him by devoting about twelve hours a day to him,' Henry confided to Philip Sheppard. 'His general peculiarities have not decreased with age . . . Good god, he is coming. In haste.'[110]

A *summa theologica* of the Oxford School of Ecological Genetics, Henry's book was the long-awaited consummation of a thirty-year research programme. It began loftily, proclaiming the founding of a new science. 'It is a surprising fact that evolution, the fundamental concept of biology, has rarely been studied in wild populations by the fundamental techniques of science, those of observation and experiment. Consequently, the process has seldom been detected and analysed in action. However, I have for many years attempted to remedy [that situation] by a method . . . which combines fieldwork and laboratory genetics.' The book focused rather parochially on the exploits of 'The Oxford School', with six chapters devoted to butterflies and moths (*Maniola, Papilio, Panaxia*, etc.), one to snails (*Cepaea*), one to plants. Naturally, industrial melanism figured prominently. 'It now provides one of the most complete pictures of evolution in progress that has so far been observed: an achievement due to the originality and enthusiasm of H.B.D. Kettlewell. His own researches have demonstrated to an outstanding degree the value of combining laboratory genetics with accurate and controlled observations in the field, while he has organized on an extensive scale the supply of information from moth collectors scattered throughout Britain.' Anyone reading the book would have thought E.B. Ford admired H.B.D. Kettlewell almost excessively.

Ecological Genetics was received as one of the monumental works in the field, and Henry would continue to produce new editions over the years, never straying far from his handful of core themes. (One colleague who was assigned to review it in one of its later incarnations complained to a friend: 'What to say when he keeps writing the same book?') He had been dictating the text over several years, and its seamless prose was always praised by critics, but when I opened the book I was surprised to find the polished, serpentine Fordian sentences curiously lifeless. Everyone who ever heard Ford lecture has said that he was magnificent, and I have caught a bit of that magic on an audiotape he made with Bernard in the 1960s, but some reverse alchemy must have obtruded between the spoken and the written word, for E.B. Ford, the famous raconteur and polymath, arrives on the page dry and etiolated, like certain fish that lose their beautiful colours on dry land. The antithesis of the apparently dull person who pours into his poems or novel an unsuspected and opulent inner life, Henry kept his interior life well concealed.

In a review in *Science*, his friend Dobzhansky noted approvingly: 'Ford is consistently and rigorously a Darwinist and a selectionist. Much of the evidence available on the action of natural selection in wild populations of higher organisms, including some of the most direct and conclusive evidence, has been brought to light by the Oxford School. Perhaps the most spectacular is industrial melanism . . .' The fact that the 'most spectacular' case made its full literary debut in E.B. Ford's masterwork – just as earlier it had appeared in his *Moths* before it was published in a scientific journal by Bernard – subtly instilled the notion that Ford was the brains behind the experiments.

In *his* review J.B.S. Haldane[111] remarked that while *Ecological Genetics* was the best book Ford had written, 'much of it is polemical and Ford would be superhuman if he were right in every controversy . . .' He agreed with many of Ford's opinions but took issue with a passage about industrial melanism in which Ford suggested that he had discovered 'far greater selection pressures' than previously believed by 'direct observation'. Not so, Haldane objected; it was he himself who said that the melanic phenotype had a 50 per cent advantage ('not 30% as Ford states') over the typical, and he made this discovery through calculations, not direct observation. Kettlewell later 'confirmed this order of magnitude by direct observation. But until we know the mean ages at which these moths copulate and lay their eggs, the exact selective advantage cannot be calculated.'

Haldane had made this point before. Having estimated the selective advantage of the melanic form of the peppered moth in 1924, he was inspired by Bernard's data to do an update in 1957. Since Bernard's recapture figures for Birmingham showed the percentage of typical recaptures (13.0 per cent) as roughly half that of the *carbonaria* recaptures (27.5 per cent), Haldane calculated the selective advantage of the melanic form in Birmingham at 50 per cent. This figure would become immortalized in countless books and articles, yet it is based only on a single predation experiment, and Haldane himself warned of its limitations. 'Of course, this does not really answer the question. One would want to know what the extra mortality of T [typical] is per day, and also what fraction of eggs are laid on the various days of life, and what fraction of successful copulations by males take place on the various days of life. Such figures, especially the last, are hard to get, even with *Drosophila*.'[112]

Bernard and Haldane, who were both larger-than-life personalities with booming voices and mischievous senses of humour, had formed a warm and jocular friendship, and Haldane often functioned unofficially as Bernard's mathematics guru. In the 1930s Haldane had declared himself a Marxist, as many intellectuals of his generation did, but later became disillusioned with the party line and with the rise of a new Marxist form of Lamarckism – Lysenkoism – in the Soviet Union. In 1957, the same year he recalculated the rise of industrial melanism, he moved to India, where he took Indian citizenship and headed the Genetics and Biometry Laboratory in Orissa. Now he usually appeared, even in Britain, with his large frame swathed in a dhoti. All his life he enjoyed rattling other people's cages, and was once moved to prankishly recommend a Hindu associate as a field assistant to Bernard and Henry, warning them that he is 'not merely a vegetarian . . . but might vomit if he had to wash dishes which had just been used for meat'. Clearly, Haldane would have known well that these dietary restrictions would have been an impossibility while camping in Northern Scotland.

While Haldane, with his expansive, *épater-les-bourgeois* ways, got along famously with Bernard, he was the temperamental opposite of E.B. Ford, who went to great pains to avoid him whenever he came to Oxford. On one occasion Haldane was attending a ceremony near the genetics offices, and Ford, loath to run into him, was holed up with several of his graduate students. 'We were putting off going to Cothill – it was July. Robert Creed and I were tipped off by Ford to keep an eye out to see when Haldane left the physiology building,' David Jones remembers. 'We finally decided Haldane had left, and went outside, and we were just getting into Robert Creed's car, which was a Brooklands Riley competition car, when Haldane and the physiology party appeared.'

'Ah Henry, I wanted to talk to you,' said Haldane. Ford replied, from the tiny open two-seater in which he was seated not more than four inches off the road: 'I am so glad to see you, Jack. I would be delighted to talk to you, but we are very busy. In fact, as you see, we are so busy that we have to use a motor racing car to get about.'[113]

9

The cost of selection

It is possible to read in the inventory of Bernard's physical complaints at any given period the state of his psyche. His hypochondria, made all the more excruciating by his medical training, was a running family joke. 'I used to bring him breakfast in bed,' said his son, David, 'and he'd be sitting there reading a medical journal. He'd say, I have the most awful headache, or stomach ache or something. And I'd say, Do you think you might have such and such disease? And he'd say, Yes! How did you know? I'd say, That was the last article you read in the medical journal.'

The same overactive imagination that had tormented him all his life with worst-case scenarios – plane crashes, nuclear war, terminal illnesses – may have also predisposed Bernard to fill in the blanks in his scientific work, leaping to certain conclusions while ignoring equally plausible alternatives. Henry complained to friends that Bernard always saw things in black and white, and colleagues noted that he could rarely be diverted from a certain train of thought once he had embraced it. 'He was the best naturalist I have ever met,' R.J. Berry recalled, 'and almost the worst professional scientist I have ever known. Writing papers with him was traumatic; as an experienced clinician he made rapid diagnoses and refused to be diverted by what he regarded as irrelevant evidence.'[114]

As a medical man, he understood something about psychosomatic illnesses. The element of his Darwin biography that intrigued him most was the mystery of Darwin's lifelong malaise, which some scholars dismissed as hypochondria and others attributed to an infectious disease contracted during his travels in South America. On his best days Darwin could be productive for only two or three hours; on his worst days he was incapacitated. During a visit to the Darwin estate in Down, Bernard 'came across reams of paper which he wrote daily between 1849–1854, when he would have been about 40, giving his exact health for each day and each night,' he wrote to Cyril Clarke. 'He had a [code] which I have not yet cracked, though I think that "sh" must refer to "shivering" . . . but there are heaps of hieroglyphics and the word "fit" occurs almost daily . . .' Bernard deemed it an 'incredible piece of documentary evidence of extreme hypochondria'.[115]

While frequently a *malade imaginaire*, Bernard also did suffer more than the usual run of real ailments. He had frequent flare-ups of lumbago, and was always throwing his back out by overexertion in the field. He suffered recurrent bouts of bronchitis, pneumonia, pleurisy and flu, especially when he travelled or overdid, which was often. He was prone to disquieting heart symptoms, fibrillations and flutterings. While on holiday in Spain with Hazel in 1965, he was stricken with painful esophageal spasms, which a barium swallow diagnosed as a 'fairly severe hiatus hernia along with regurgitation oesophagitis'. Cyril Clarke took X-rays in the spring of 1966 and informed him that his diaphragmatic hernia was inoperable, and that, in Bernard's words, 'any emotional stress-strain or worry brings it on'. By this time he was thoroughly miserable, with 'spasm in practically every portion of my anatomy due to the "psychologics".'

Part of his stress was due to the struggle to bring his sprawling breeding data up to date on his 'dominance breakdown–buildup' experiments. These painstaking, interminable breeding studies were inspired by the belief that Ronald Fisher had impressed deeply on the Oxford School that dominance was not the fixed property of a particular allele but depended on modifying genes in the overall 'genetic background' of the organism. Crossing domesticated poultry with wild jungle fowl years before, Fisher had achieved 'dominance breakdown', undoing the dominance of certain genes after a few generations and getting chickens with a new range of traits. Henry himself had done something similar with the currant moth *Abraxis grossulariata* and with *Panaxia*; Philip Sheppard had dazzled everyone with his work on *Papilio*, the African swallowtail butterfly; and now Bernard drove himself to achieve a comparable success in *Biston betularia*.

The melanic allele in *Biston* was fully dominant, so a heterozygote looked just like a homozygote. If you crossed a British melanic and a British typical you would get jet-black melanic offspring and pale typicals, no intermediates. But what if you introduced the British *carbonaria* gene into a foreign gene pool, one without the same complex of 'modifying genes'? For years Bernard selectively bred Birmingham melanics to *Biston cognataria*, the North American peppered moth. He took the melanic offspring of each pairing and back-crossed them with

North American typicals, and after three generations reported a full range of intermediate types – black but peppered with white dots, grey-black with pale bands, et cetera. The dominance of the Birmingham melanic had 'broken down', in the parlance, because the North American moths did not possess the same 'modifiers' in their genome. This implied that the dominance had been 'built up' in the first place in the British species by the gradual accretion of modifying genes.

The next step was to unravel his breakdown work, transferring the melanic allele back into its normal British 'genetic background'. He did this by mating a melanic from the third generation of the North American back-cross broods with a typical peppered moth from Birmingham – and, voilà, after a few generations the offspring consisted once again of binary phenotypes, melanic and typical. (Bernard did another series of back-cross experiments between Birmingham melanics and typicals from Cornwall, a gene pool in which melanics had not appeared, and got less dramatic results.) Nosing around nineteenth-century collections, including one at Oxford's Hope Department, he was excited to notice that some early British *carbonaria* were *not* inky black but speckled with white dots like some of the intermediates in the back-cross broods. Perhaps, he proposed, the melanic allele had not always been fully dominant in Britain, but had been 'built up' in the past through the slow accumulation of modifier genes. This was a problematic notion given the recentness of industrial melanism, and Bernard's attempts to wiggle around it were never entirely successful. In any case, after managing to summon up the muse of science-journal dullness, Bernard published his results in *Science*.[116]

Posterity has not been kind to Bernard's dominance breakdown–buildup experiments. 'West crossed British *carbonaria* with typical from the Appalachians where the melanic had a frequency of only about 1% and observed no breakdown in dominance,' R.J. Berry observed in a 1990 paper. 'Likewise, Mikkola . . . found no breakdown in crosses between *carbonaria* from Liverpool and *typica* from Finland.' Berry also noted that moths like Kettlewell's 'ancient' *carbonaria*, which he believed were a lighter, not fully dominant form of melanic, are regularly found in modern samples from the Liverpool and Birmingham areas.[117]

The worst of Bernard's 'psychologics' were centred not on his scientific struggles, however, but on the ongoing tragedy of Dawn, whose latest 'ghastly affair' (in Henry's words) was unfolding in 1966, just as Bernard was finally sitting down to write his long-postponed book on melanism. Bernard and Hazel had held their breath and hoped that marriage, in 1961, would settle their wayward daughter down, although they had been realistic enough to drop veiled hints to her schoolmaster husband, of whom they were very fond, about what he was taking on. Unfortunately, domesticity and the birth of two children had not exorcized Dawn's inner demons, and by 1966 she was caught in bed with two boys at the distinguished public school where her husband was a master. Her husband was asked to leave amid a humiliating public scandal, the marriage was understandably coming unglued, and in the course of trying to straighten out his daughter's life, Bernard discovered that a man he had regarded as a trusted colleague and friend had been another of Dawn's lovers. A psychiatrist offered the opinion that Dawn was 'psychopathic', Bernard wrote sadly to a friend, 'though I do not believe this.'

'There was always this background of dismay,' recalls Miriam Rothschild. 'He would tear his hair out, saying "I am in a dreadful state."' When Hazel and Bernard took refuge one weekend at Miriam Rothschild's house, she tried to bolster their spirits as best she could, and someone who saw him just after Easter said he was his usual ebullient self. After being collegeless all these years – a fate akin, at Oxford, to being a man without a country – he was made a fellow of the newly founded Iffley College, whose first president was Isaiah Berlin; later it would be renamed Wolfson College.[118]

In April 1966 Bernard accepted an invitation to attend a symposium on 'Mathematical Challenges to the Neo-Darwinian Interpretation of Evolution', at the Wistar Institute in Philadelphia. Thus he happened to be present at one of the first public (and quite ill-tempered) challenges to the neo-Darwinian worldview. The chairman, the Nobel-prize-winning biologist Sir Peter Medawar, opened the meeting by saying: 'The immediate cause of this conference is a pretty widespread dissatisfaction about what has come to be thought of as the accepted evolutionary theory in the English-speaking world . . .' There was a statistical crisis. Having worked out the mathematical consequences of key evolutionary theories in the latest mainframe computers, some mathematicians were expressing open astonishment at the evolutionists' optimism about what could be achieved by random mutations. And many people there seemed to agree with Conrad H. Waddington, professor of biology at Edinburgh University and a person deplored by E.B. Ford, that the central Darwinian proposition was an empty tautology. 'Natural selection, which was first considered as though it were a hypothesis that was in need of experimental or observational confirmation,' Waddington insisted, 'turns out on closer inspection to be a tautology, a statement of an inevitable although previously unrecognized relation. It states that the fittest individuals in a population (defined as those who leave the most offspring) will leave the most offspring. Once the statement is made, its truth is apparent.' The survivors survive, in other words.[119] Clearly, much of this passed over Bernard's head. 'The Darwinians talked a completely different language from the "computer boys" and vice-versa, but it was great fun even though it nearly came to blows at times,' he wrote to a friend.

In June, while attending the Mendel centenary in Brno, he caught a larva of *Apatale aceris*, which proceeded to crawl into the glass containing his false teeth and pupate there as Bernard slept in a hotel room in Prague. He brought the pupa back to England and waited for it to hatch. That autumn he was thrilled to find rare Lepidoptera from southern Europe in his moth traps, including a male death's head hawkmoth, *Acherontia atropos*, whose markings eerily resemble a human skull. Bernard recorded its weird, squeaky sounds, and when he played the recording back while lying in his bath, a mouse 'went crazy and tore around the room as soon as the stridulations struck up'. This sort of adventure was typical of Bernard.

Apart from his entomological escapades, Bernard was in the doldrums as 1966 passed into 1967. His friend Julian Huxley was in a nursing home receiving electroshock treatment for one of his periodic depressions, and Bernard noted with astonishment that for the first time in his life Huxley's handwriting was legible. His other ally, Haldane, had succumbed to cancer of the colon in December 1964. When Bernard had visited him in hospital shortly after his diagnosis, he found his friend in high spirits, composing a long poem that began: 'My rectum is a loss to me/ But thanks to my colostomy . . .'[120] Bernard's own aches and pains were getting no better, as his helpless anguish at Dawn's unravelling was compounded by the interminable ordeal of trying to write his '*magnum opus*', as he always called it. 'I am finding it very heavy going; I had looked forward to writing it but in a strange way all enthusiasm is gone,' he wrote glumly to a friend. He would lock himself in his study at Steeple Barton every morning at ten o'clock to wrestle unhappily with his demons. 'He'd get into a state of terrible worry, and he didn't know how to put it down,' his son recalled. 'It drove my mother mad.'

All the while Henry never ceased reminding him that he simply *must* write his book. 'I do hope you are really able to settle down to the book: the most important thing you could do now,' Henry was writing Bernard from the University of Minnesota, which he described as 'brick and of revolting appearance. There are 35,000 undergraduates, who pass through it like herds of animals . . .'[121] Touring American universities, Henry was continually amazed by such American phenomena as seminars (a 'frightful form of entertainment'), and the 'growling voices of American men and the quacking of American women' offended his sensibilities as deeply as the swarming undergraduates. But the deep snowdrifts of North Dakota, the immense emptiness of the Great Plains, the lush woods of North Carolina enchanted him, and, like Bernard, Henry always sounded happiest when he was far from Oxford. The strain of being Henry, of maintaining the impeccable act, must sometimes have taken its toll.

Earning a few paragraphs in a general biology textbook is a bit like getting into the highest stratum of Dante's Paradise, and thus when Kettlewell's experiments began to filter into the textbooks in the mid-1960s it signified that they had become the equivalent of classics. From this time on, biology students throughout the world would absorb the story of the light and dark moths in England that were living proof of Darwin's theory of natural selection. The twin black-and-white photographs, set side by side, suggest a parable, or one of those Victorian cautionary tales in which a good child is rewarded and a bad child meets an unfortunate fate. In the first, a black moth with spread wings is starkly silhouetted against a tree trunk encrusted with lichen. Only an intruding arrow alerts us to the presence of a second, light-coloured peppered moth, invisible against the lichen. The second photograph shows us a dark, sooty oak trunk adorned with a speckled peppered moth and its unseen *carbonaria* twin, denoted with an arrow. The arresting photographs guaranteed that the experiment would be imprinted on a generation of young minds. After briefly tracing the development of industrial melanism, and asking rhetorically 'Why the change?', an American textbook published in 1966 continues:

The answer is almost self-evident from the photographs shown in Figure {1}. In A we see a tree trunk of the sort found in rural England far from industrial centers: lichens covering the oak tree give it a variegated surface against which the lightly peppered moth is hard to see; the black form stands out prominently. By contrast, on trees growing in industrial areas, the lichens are killed and the trunk is blackened by soot; on such a tree it is the black moth that is protectively colored, the light moth standing out like a sore thumb. Birds that prey on the moths have been observed and photographed catching moths, and it has been proved that they bring about differential mortality favoring the survival of the light forms in unpolluted woods and the dark forms in industrially blackened woods.[122]

This paragraph was typical: black and white, without a shade of grey. Crypsis, lichens and selective bird predation were the only elements of the story, no ifs, ands or buts. All of it had been 'proved'. E.B. Ford's mission, conceived in 1927, to prove scientifically the operation of natural selection in nature had succeeded better than even he had dreamed.

There was a nagging question, however. By what strange coincidence had a mutation happened to produce melanic forms in a hundred species in the British Isles alone just in time for the Industrial Revolution? This is a case of the general problem known as 'preadaptation'. According to neo-Darwinian theory, a mutation cannot be other than random (if it were not, we would fall into theories of Lamarckian directed evolution), and it must appear *before* natural selection can seize it and act on it. E.B. Ford had hypothesized that melanism had arisen many times in the distant past, that a melanic mutant might have been preserved for one evolutionary reason and then spread widely for another, and Bernard tried to find evidence for this idea. If there were natural reservoirs of melanic genes, he thought, the right moths would be available when industrial soot coated the landscape.

To solve this puzzle, he turned his attention to species that had turned melanic in rural environments, far from industrial grime. It was not just that he was looking for excuses to be far from Oxford, though he was. He was also honing his thesis that melanism was a 'recurring necessity', arising wherever there was a darkened background against which a dark-coloured moth would be cryptic. (According to this hypothesis, industrial pollution would simply be the latest necessity.) Each August for many years, Bernard and a shifting group of helpers would take off for the windswept, boulder-strewn shores of Unst, the northernmost of the Shetland Islands, to do predation experiments on *Amathes glareosa*, the autumnal rustic, a pale grey moth with a blackened form that was plentiful in the far north. These melanic moths had thrived, he believed, due to the selective pressures of flying in sub-Arctic twilight against black, peaty soil. Gulls, he thought, were the agents of selection.

On Unst the scientists stayed in a lobster fisherman's hut in the corrugated-tin shanty town of Baltasound, where Bernard, who was famously mad for lobster, would 'dissect them down to the last antenna,' recalls James Cadbury,[123] the son of Bernard's old friends, who worked as his assistant. The aurora borealis put on a display at night, gulls wheeled and dived among fjordlike cliffs, and Russian trawlers with huge antennae plied the frigid waters, playing Cold War chicken with the local American radar station. The light form of the moth was more numerous in the south, the dark one in the north, and the north–south transect formed a gradient, or 'cline', which was believed to advertise a situation of strong natural selection. In order to have enough of both forms in both places for their selection experiments, Bernard and his helpers ran a moth shuttle service involving three buses and two ferry boats each day. 'My job,' says R.J. (Sam) Berry, 'was to catch the white ones at the southern end and then get on the bus and release them in the north, while James Cadbury would be catching the dark forms in the north and releasing them in the south.' Haldane wrote a paper about the cline, as did Bernard and Sam Berry, and there was much academic discussion of it, but the situation turned out to be far more complicated than anticipated, and, as Berry put it, 'We never got to the real story.' The entomologists did make a vivid impression on the locals, for the island's annual Uphalla festival began to incorporate the doings of the 'Mothy Men'.[124]

Berry joined Bernard, Cadbury and others, in fieldwork on another rural moth species with a melanic form – the oak eggar, *Lasiocampa quercus*. Bernard sampled oak eggars frequently in Scotland, Yorkshire and the Lancashire coast, carting back quantities to raise and breed in Oxford. While on a date with James Cadbury, Kate Davies recalls hearing from the back of the van the sound of 'armies of oak eggar caterpillars munching their way through vegetation . . . It's quite noisy, like the sound inside your head when you're eating lettuce.' Bernard hoped his study of rural melanism might plug the holes in his theory of the evolution of dominance, as well as the problem of melanism's origin, but in the end it would solve neither.

Meanwhile, an important new chapter was being added to the peppered moth story. After several prize cattle died from the effects of smog at a national cattle show and opera perfomances at London's Sadler's Wells Theatre were cancelled because the audience could not see the stage, clean air acts began to be introduced in the UK in 1956. A few years later, a shift in the balance of the different morphs of the peppered moth was being observed in a few areas near the new 'smokeless zones'. Previously, during a visit to Liverpool, Bernard had talked his old friend Cyril Clarke into trapping peppered moths, and the habit had taken hold. Since 1959 Clarke had run his moth trap regularly in his garden in Caldy Common, on the Wirral Peninsula, 18 miles from Liverpool, and he was able to report that the frequency of melanics had *declined* rather noticeably since 1959. A smokeless zone lay to the east. Clarke and his collaborator, Philip Sheppard, wrote a paper announcing their results in 1966.

The evolution of melanism had been documented only spottily, and Ford had always bemoaned the 'lost scientific opportunity', but biologists realized with growing excitement that they could now monitor melanism's decline *as it happened*. Melanism in reverse should strengthen the case for the classic model of industrial melanism; it was, in a sense, the other half of the experiment, which makes Bernard's stubborn resistance to it so puzzling. Not only did he take issue with Philip's and Cyril's data, but when a sharp increase in typicals was documented in Manchester by Laurence Cook and Jim Bishop a few years later, Bernard, in Sheppard's words, 'tried to throw doubt on the evidence by pointing out that *insularia* and typical may be confused and therefore the change is not real'.[125] It was real, but privately Sheppard himself seems to have been troubled 'that the change in frequency appeared far too rapid for a change in gene frequency,' as he wrote to Bernard as his paper was coming out. 'It virtually occurred in one year, or at the most two, which would mean changes in selective value of a very high magnitude.'[126] In other words, such a sharp drop in frequency was asking natural selection to work awfully fast.

In Caldy Common, which had birches, sycamores and oaks, Clarke and Sheppard performed a predation experiment similar to Bernard's. They wanted to confirm the two key elements of Bernard's work: that the colouring of the morphs really did provide crypsis, and that there was differential predation by birds. They avoided the inconvenience of getting moths to pupate at just the right moment by using dead, frozen moths, glued onto tree trunks 'in an attitude as similar as possible to that of a normal resting moth'. Because they suspected that Bernard had created an artificial situation with the high density of moths he used (about four per tree), they set out only four moths per day, one per tree. Three predation experiments were carried out: one with both forms placed on the darkest patch of trunk; another with both forms placed so as to be the most inconspicuous; a third with both forms placed as high as possible on the tree trunk. In a general way, their results seemed consistent with Bernard's, although they found, surprisingly, that *carbonaria* was at an advantage on the white bark of birch. 'It seems likely that it is mistaken for the black rough areas,' Philip wrote Bernard.

Clarke and Sheppard had also conducted a survey in Caldy and at eleven sites in North Wales chosen, it was said, for their proximity to fly-fishing sites and pubs favoured by Sheppard. This first concentrated sampling of peppered moth frequencies in a single locale posed potential difficulties for the standard explanation of industrial melanism. Given the decline in *carbonaria* frequency in Clarke's garden in Caldy – a drop of 2 per cent between 1962 and 1964 – the light-coloured typical morph should have had an *advantage* of about 23 per cent, Clarke and Sheppard calculated, but in their predation experiments typicals still had a selective *disadvantage* of about 20 per cent. It didn't add up. 'The decrease in the frequency of *carbonaria* therefore requires some compensating selective disadvantage,' they concluded. 'It seems very probable that this selection results from some "non-visual" disadvantage of the homozygotes . . .' (This will not be the last time a mysterious 'other' factor is invoked, like a *deus ex machina*, to save the theory.)[127]

Bernard was prickly and defensive about his friends' experiment. The rough draft that Philip sent to him was returned riddled with marginal notes and long-winded, often irrelevant objections, and Bernard always opposed the use of dead moths 'because dead specimens cannot compare with the perfect crypsis obtained by living insects'.

His professional touchiness was likely exacerbated by his worsening physical complaints, or perhaps it was the other way around. By 1967, his diaphragmatic hernia was getting no better, his 'plagues' were more frequent, and he was feeling increasingly moribund. Henry, who always took meticulous care of himself, had little sympathy for Bernard's serial maladies, even less for what he saw as his laziness. Henry was insisting on seeing Bernard's chapters as they were churned out, and what he saw did not please him. Philip and Henry were frequently wringing their hands by post, Philip lamenting: 'I am afraid I made [my comments] rather discouraging but I feel the chapter [8] is even worse than chapter 7,' and Henry writing back: 'It sounds as if it will be pretty grim. I must do my best.'[128]

The incessant criticism was driving Bernard to distraction, and his anguish became translated into acute abdominal pain, which turned out to be diverticulitis. Miriam Rothschild tried to bolster his spirits: 'Don't be put off by Henry. You write much better than he does. Henry explains everything very well but he can't write English for toffee.' Rothschild, among others, suspected that Henry was deliberately hazing Bernard, that the continual rewrites were meant to have precisely the demoralizing effect they did, causing Bernard an infinite number of delays.

His private sorrows did not visibly impair his talents as an expansive and genial host. A family of Kettlewells from Ohio was visiting Oxford that spring, and knowing that the 'most famous Kettlewell in the world' lived nearby, rang up Bernard on the spur of the moment and was immediately invited to Steeple Barton. Jim Kettlewell, then nineteen and in the army, remembers Bernard's large hands 'pulling moths out of his trap one after another. You could tell he'd done it a thousand times. He would reject certain ones if they were not the right gender or variety or to show us something he wanted us to learn.' Insisting on a group photo, Bernard rummaged through a trunk in a closet and dramatically produced an assortment of funny hats from his travels. 'I remember him looking at each of us before deciding which hat he would give us, as if it really mattered,' recalls Jim Kettlewell. 'He was being gracious to perfect strangers, who had been bold enough to intrude into his life one afternoon. Of course, it was evident that he was very flattered by all the attention.'[129]

It was 1968, a tumultuous year. Catholics in Northern Ireland began militating for full equality, and a high-rise tower block in the slums of East London collapsed, revealing the shoddiness of the new urban housing. By early summer, protests against the Vietnam War were bringing down an American president, Janis Joplin and Big Brother and the Holding Company were recording 'Take Another Little Piece of My Heart', and a brutal bloodbath would spill over into the streets of Chicago. Undeterred by widespread urban riots following the assassination of Martin Luther King, Jr in April, Bernard crossed the Atlantic in June to attend the Lepidopterists Society's annual convention in Washington, DC. After three of the gathering were mugged and beaten in the streets, Bernard couldn't get away from the American capital fast enough. His spirits were buoyed by a tour of the New Jersey Pine Barrens, exactly the sort of ancient coniferous forest that he believed would harbour nonindustrial melanic Lepidoptera. The first New World industrial melanic had appeared about fifty years after the first Manchester *carbonaria*, and over the ensuing years melanics appeared in many species including *Biston betularia*'s American cousin, *Biston cognataria*. Its black form, called *swettaria*, had first been sighted in Philadelphia in 1906. The sight of the vast tracts of pitch pine in central New Jersey had a Eureka effect on Bernard's mind: 'Whereas in Britain the whole of our indigenous Pine forests . . . have long since been cut down, your Pitch Pine forests are virtually indestructible . . .' he wrote to an American collector friend. '[They] must be of great antiquity and may have been able to maintain melanic polymorphisms throughout the ages, and these, of course, would have acted as ready reservoirs in North America. By contrast, we in Britain have had to await an actual mutation to take place and hence even after 150 years or more, new melanic forms are still occurring. This, of course, is pure theory.'[130]

The New World was on Henry's mind, too, for a different reason. He had always worked very hard to stir up American and Canadian enthusiasm for Ecological Genetics, touring both countries at frequent intervals, but now there was a new heresy in the air. It was called the neutral theory.

Up to the early to mid-1960s, the Ford/Fisher worldview prevailed, with the consensus that every biological trait could be interpreted in the light of adaptive evolution and that almost no mutant genes were selectively neutral. In 1930 Fisher had 'proved' mathematically that selective neutrality must be exceedingly rare, since it would require a precise balance of advantage and disadvantage between two alleles; neutral genes could not take hold in a population because fluctuations in the environment, or in the genetic outfit of the organism, would quickly upset the balance.

The earlier debate had been over how much genetic variation existed in nature. The 'classical' school insisted that there were few variants, and those largely deleterious, while the 'balance school' – including Ford *et al.* – said that there was a rich supply of genetic variation on which natural selection could work. Finally, in the mid-Sixties, the question was answered. Using a new molecular technique called gel electrophoresis, scientists could identify genetic variants of certain enzymes, and they discovered that in every organism tested, from fruit flies to man, genetic variation was greater than anyone dreamed. At first Ford seized upon this as the best possible news, for lots of polymorphisms were splendid from his point of view. 'We are getting exciting results here,' he wrote Bernard in 1966, from the Devon–Cornwall border, where he was once more counting spots on *Maniola jurtina* butterflies, 'and are sending off living material [*Maniola jurtina* butterflies] to Oxford for Gilbert[131] to carry out esterase studies on it. Also to Texas by means of the flying machine. This, we hope, may tell us much.' His pleasure was shortlived, for it soon became apparent that there was *too much variation*.

Here the peppered moth plays a role. Its evolutionary rate was cited in an influential 1957 paper by Haldane, called 'The

Cost of Natural Selection', in which he argued that whenever one allele is replaced by another allele by natural selection, there is a cost in lives. When the *carbonaria* allele of *Biston betularia* was at a frequency of 70 per cent, one-third of the typical homozygotes, and 16 per cent of the population, were being destroyed per generation. To bring a favourable gene to a high frequency inevitably requires the deaths of many individuals with the unfavourable gene. Suppose that 'a breeder selecting for higher milk yield in cattle must prevent the lower-yielding individuals from breeding, perhaps by slaughtering them', evolutionary biologist Douglas Futuyma explains. 'If the fraction killed in each generation is small, the herd remains large but genetic change is slow. If the fraction killed is large, genetic change is fast but the herd size is reduced . . .' The stronger the selection, the fewer the survivors.[132]

Haldane's calculations showed that it would take 300 generations, on average, to 'pay' the cost of substituting just one trait. It was possible to run the numbers and conclude that many higher vertebrate species could not plausibly evolve – substitute enough genes – in the available time and therefore macro-evolution was mathematically impossible. Whether Haldane's Dilemma, as it became known, has ever been convincingly solved is still being debated, and there is a school of thought that insists that since our presumed divergence from ape lines there has been nowhere near enough time to substitute the required number of genes.[133]

The 'cost of selection' reared its head again when the first techniques of molecular population genetics were devised in the mid-1960s. By grinding up the tissues of organisms and running them through a medium (usually potato starch) scientists could separate different forms of the proteins (enzymes) in the presence of an electrical field. Allozyme electrophoresis, as it was called, provided a quick and easy method of measuring genetic diversity at the molecular level, and in 1966 H. Harris was identifying human enzyme polymorphisms, while Richard Lewontin and J.L. Hubby applied the technique to *Drosophila*.[134] Electrophoresis revealed enormous genetic variation at the molecular level. A startling 30 per cent of the *Drosophila* genome, or about 2,000 genetic loci, was polymorphic. If natural selection had to maintain so many polymorphisms by eliminating less fit individuals, wouldn't the 'cost' be prohibitive? That was the gist of a landmark 1969 paper in *Science* by Jack Lester King and Thomas H. Jukes entitled 'Non-Darwinian Evolution', and their argument was developed fully into 'neutral theory' by Motoo Kimura and others. At the level of DNA or of the polypeptides coded for by genes, the theory holds, there are many variants that do not differ in fitness; they are selectively neutral, their frequencies determined largely by genetic drift and gene flow. (Today, virtually everyone agrees on this point, though there are differing opinions about what it all means.)[135]

Henry disagreed violently, of course, and cited scriptural authority – that is, Fisher's papers. 'I see no difficulty in having the large amount of polymorphism detected by Lewontin, Harris, etc., maintained and arising by selection,' Henry ranted to Dobzhansky. 'There is no dilemma, since the difficulty raised by Haldane's paper is based on an error . . . It is the paper of one who certainly at the time he wrote it was looking at the matter as a mathematician, not a Geneticist . . .'[136] It was typical of Henry to dismiss views he did not like by labelling the author an imbecile, and he began firing off diatribes to everyone in evolutionary biology. To his mind it just wasn't possible that molecular evolution could be largely random, or 'nonDarwinian'.[137] After briefly luring Henry with the promise of identifying his beloved polymorphisms with undreamt-of specificity, the molecular tools were turning out to be instruments of the devil, and the lovely deterministic laws of 'ecological genetics' were once more being besieged by the forces of randomness. The neutral theory was heading for a head-on collision with 'the English view of random drift as lunar green cheese', as John Turner would put it.[138]

E.B. Ford was used to being right; in fact he had *always* been right, and had managed to construct around himself a changeless universe in which he was virtually never challenged.

In 1958, due in large part to Miriam Rothschild's behind-the-scenes promoting, Henry had been made a fellow of All Souls College, a college so rarefied that it had (and still has) no undergraduates. Founded as a chapel to pray for the souls of soldiers killed in the Hundred Years War, All Souls presents an austere face to the world, even pricklier and more pinnacled than other Oxford colleges, and, lacking sprawling undergraduates and rows of bicycles, its immaculate quadrangle has a deserted look. Even the porter's lodge seems less welcoming than at other colleges and generally wears its 'closed' sign. The impression of a secret society where hush-hush business might be conducted far from the gaping *hoi polloi* — in fact, Neville Chamberlain hatched his infamous appeasement plan at All Souls in 1938 — would have appealed to Henry's love of intrigue. All Souls is a college made up of fellows, some of whom are Oxford dons and others important men of the world who appear only on weekends. In Henry's day it was for men only, and Henry would do everything in his power to keep it that way.

Henry always claimed to be the first scientist elected to All Souls since Christopher Wren, and was fond of referring to 'my immediate predecessor, Sir Christopher Wren', which was not precisely true. His repertoire of lapidary stories and recondite jokes became a fixture at High Table, where he 'might be lured by an utterance of a line of verse into continuing the quotation,' recalled Bryan Wilson, an All Souls fellow. 'He could talk easily about Ruskin's ideas of art; about volcanoes; the later years of Napoleon on St Helena; or the court life and family affairs of Queen Victoria, who was by way of being – if I may put it so – one of his "heroes" – for his view of the world did not readily admit of heroines. His knowledge of heraldry was extensive and peculiar . . .'[139]

All Souls was an artificial paradise, like Yeats's golden Byzantium. Within its stone walls lay sumptuous apartments, Oriental carpets, mahogany furniture, priceless paintings, fine sherry in antique silver decanters. In a senior common room one of a handful of original copies of *The Seven Pillars of Wisdom* by All Souls fellow T.E. Lawrence was enshrined in a glass case. Like all Oxford colleges, the college had a number of prized posts, such as Bursar, Senior Dean, Warden, in the pursuit of which ambitious dons were said to mount campaigns of high intrigue. Henry's sometime friend John Sparrow was known to have schemed elaborately to become Warden of All Souls in 1952 and to keep the post for twenty-five years; and although it is not reported how Henry might have manoeuvred behind the scenes to become the Senior Dean twice, he clearly relished his position and all the High Church pageantry that came with it.

It was Henry's seigneurial custom to invite some of the young men of his department to dine with him on certain evenings at All Souls. After a rather costly sherry in his rooms, or in a senior common room, they would proceed to High Table, where a formidable phalanx of fellows, often including a cabinet secretary or two, would sit at a single long table to be served several courses by deferential retainers. A guest could expect to be interrogated, and he had better have a supply of erudition and facts at his fingertips. 'As a nervous graduate student it's very obvious that you're being tested out for your intelligence and your wit,' recalls John Turner.

After dinner everyone would repair to another room for dessert and coffee, observing the elaborate table rules, for instance, that the wines must be passed clockwise and the ship-shaped silver decanter containing the port must precede the claret jug, which was *never* allowed to overtake it. 'It was a social museum,' recalls Bryan Clarke. 'I was invited to dinner by Henry, along with a friend, and we were seated next to A.L. Rowse [the famous Elizabethan scholar who discovered the identity of the "dark lady" of the Shakespearean sonnets]. I said I thought it wasn't a good idea to stay in Oxford all one's life, because one got divorced from reality, and Rowse said, "Oh, dear boy, reality is *so boring*." Henry lost interest in our company, and in the interval between dinner and dessert and coffee in the other room, he darted down a side corridor and disappeared completely. I wrote him a thank you note the next day, saying, "So sorry we missed you." He wrote back, "You completely disappeared. I have never known anything like it in my life."'

In 1976 William Provine, of Cornell, a biologist and an historian of the Synthesis period, was in Oxford interviewing Henry about the Sewall Wright–Ronald Fisher controversy (Henry served him high tea but did not permit him to take notes) and was invited to All Souls as his guest. 'He sat me next to this important government official, who asked what I was doing research on. At that time I was doing research on hereditary mental differences between human races, and he asked my view. I said, "I don't think you can tell anything from current data. I don't think it's a scientific issue." He stopped the whole table and announced to everyone, "Mr Provine doesn't think there are any hereditary differences between human races or between the sexes." The table fell silent and then Henry Ford said, "Oh, the evidence indicates the opposite in both cases," and everyone murmured assent to that. Then the man said, "And who is going to win the election in the fall?" I said, "Jimmy Carter's probably going to win." He announced this to everyone in the room, and there were peals of laughter at this bizarre idea that Jimmy Carter would win.

'You couldn't have made me go again,' he says. 'It was just a huge dose of high-falutin' crap and I didn't enjoy it one bit.'

Self-preoccupied and undisturbed by common human experience, Henry managed to keep the real world at bay and to surround himself only with admiring reflections. He travelled widely and frequently, to Italy, North America, the Middle East and Australia, but wherever he went, whatever he saw and heard was filtered through a constellation of *idées fixes* that had more or less solidified in the 1930s. Thus Henry Ford's view of evolution did not admit of any alteration, and he dealt with the emerging molecular realities by denying them. Visiting Canada's Maritime Provinces in March of 1968, he observed wishfully that 'people are mad-keen on the possibilities opened up by Ecological Genetics', and that 'the general impression there is that molecular genetics is a relatively old-fashioned outlook . . .' The next spring he quoted a Syracuse University professor as saying that 'Molecular Biology and Molecular Genetics are virtually out – and their popularity over, replaced by Ecological Genetics.' Henry's antennae were failing him now, or perhaps people knew to tell him what he wanted to hear. In fact, molecular genetics was in the ascendancy, and after all his years of proselytizing, the great E.B. Ford was presiding over a waning empire. He, however, perceiving himself at the hub of all things that mattered, could not really conceive of such a turn of events, and assumed he could turn the tide by clever politicking, as he might have done at All

Souls. When his old chum Dobzhansky showed signs of being drawn into the heresy, Henry wrote to a mutual friend: 'I do hope it is possible to get Doby to modify somewhat the things he is saying in the edition of his book he is now revising . . . He has got hold of that well-known paper by Haldane of about ten years ago . . .'[140]

The molecular menace arrived amid a sea of other troubles, for Henry's beloved cousin, Evelyn Clarke, had plunged into florid dementia – 'completely insane', in Henry's words – and more harrowing chapters were arriving every day from Bernard. (A letter also arrived from a Sheffield clairvoyant inquiring about 'the inheritance of parapsychological perception', to which Henry replied that he had 'arrived at the conclusion through my own observation, that inheritance of this type of perception may be controlled by Mendelian genes. I happen to have this myself, but have no progeny to test.') In the summer of 1969 Henry was convalescing at home after surgery for a non-malignant obstruction of the colon, which prevented him from going jurting. Bernard and Hazel had been nice about visiting him and bringing him eggs from their Steeple Barton chickens, and Henry thanked Bernard effusively for being a 'true friend' in an uncharacteristic letter in which the mask slips a bit, revealing the essential loneliness beneath. ('I can never thank you properly . . . Your visits gave me the encouragement that I needed so much.')[141] But his little makeshift family was disintegrating, as was his intellectual demesne: Mrs Clarke would be dead by Christmas, while young Kevin had emigrated to Los Angeles and would shortly sever all ties with Henry. The graduate students in his department with whom he formed bonds of friendship inevitably got their doctorates in the fullness of time and moved on to teaching positions elsewhere.

As his official retirement in 1969 drew near, Henry plotted

to pass the torch to Philip Sheppard, his scientific heir apparent, but all his behind-the-scenes gamesmanship was in vain, for the 'evil genius of molecular genetics' was on the rise. Not only did the professorship go to a molecular man, but the lectureship he had hoped to secure for his protégé Robert Creed went to a biochemical geneticist. In a low moment, he complained to Sheppard: 'I can hardly believe that at the end of my time what I have tried to do here should have been set on one side like this. It is *very bitter* to me.'[142] His anguish was Lear-like, sharper than a serpent's tooth. Ever the consummate actor, E.B. Ford continued to behave as if he were in control of the world, while he must have sensed that the future did not belong to him.

IO

The goose that laid the golden eggs

When *The Evolution of Melanism* by H.B.D. Kettlewell was published by Clarendon Press in 1973, its author had no illusions about its probable reception. He wrote to Julian Huxley: 'I think I shall flee the country for a year until the reviews are finished. The poor book was killed before it was born.'[143] In Bernard's mind it was largely his former friends, Henry and Philip, who had already 'killed' his unborn masterpiece.

For almost a decade he had suffered the torment of trying to write a book that had galloped away from him like an unruly horse. He was repeatedly undone by the fact that, after he had written up a certain set of survey data, new data would appear before he had managed to complete the book, and he was faced with the inconvenience of having to include them.

The organization of the chapters was giving him headaches, and he rejected Philip's and Henry's advice not to try to cover melanism in all species. Henry kept demanding to see chapters and then sending them back severely blue-pencilled. When at the end of his Sisyphean labours Bernard showed the completed manuscript to his boss, Henry 'told him that it was a lot of rubbish and he would have to rewrite it', according to his son, David. Distraught, Bernard sought the advice of his friends Julian Huxley and Miriam Rothschild, who were 'very helpful and urged him to go ahead and publish it as it was', according to David. This Bernard did, defying Henry.

All of these anxieties were mirrored in somatic afflictions; the ache in his heart was reflected in chest pain, heart trouble, pleurisy, double pneumonia. Bernard's heart had been fibrillating badly and he had been feeling progressively weary for several years before he was diagnosed with a '4:1 heart block' in 1971. His letters and conversations increasingly revealed a morbid preoccupation with skipped beats, 'awful tumblings', and flutterings, which medical books refer to as 'heart consciousness'. In April 1972 he was in hospital with pleurisy and pneumonia; in June he was being hospitalized once again for his 'damn silly heart condition'.[144] Low blood pressure and back pain added to his miseries. And that was *before* his book came out.

For all its defects, *The Evolution of Melanism* is a beautifully produced book, with exquisite if confusingly labelled plates showing various moths in various attitudes. It begins on a portentous note: 'Amongst all living things it has fallen to the Lepidoptera to provide evidence of the most striking evolutionary change in nature ever to be witnessed by Man.' Unfortunately, it quickly emerges that the author has become bogged down in his immense subject, losing sight of the forest for the trees. The first chapter on 'The attributes of black colouration' is soporific, and the reader must plough through several dense discussions of melanism in other species, on world melanism, and such, before finally arriving at Bernard's experiments in Chapter 8, in which his 1955 and 1956 *Heredity* papers are reproduced virtually verbatim.

Julian Huxley wrote him a nice congratulatory note. Henry's first comments were tepid and insincere: 'Certainly I do think your book looks extremely nice and the little chance I have had to examine it suggests a vast amount of interest.'[145]

In the non-academic-press world it would be peculiar indeed to assign a book review to the author's friend, yet it may not have been unusual in those days for the leading British scientific journal, *Nature*, to ask Philip Sheppard to review Bernard's book.[146] His review can best be described as 'mixed', for Philip did find a few nice things to say. 'The text of this book makes extremely fascinating reading and will fascinate the general reader concerned with the study of micro-evolution,' he wrote, and 'one of its very great merits is the large number of speculative conclusions which have never been published before . . .' But he did not try to mask his distaste for the obscuring fog of typographical errors, mislabelled drawings and legends, arithmetical mistakes, and data in appendices that clashed with other data in the text or in other appendices. There were statistical howlers and erroneous citations, and Bernard's refusal to update his earlier writing had given rise to appalling contradictions.

When he sent a pre-publication draft to Bernard – another quaint custom – Bernard saw through Philip's lukewarm praise and attacked him for failing to mention his conclusions on the natural history of melanism and on the evolution of dominance. Philip hadn't mentioned these subjects because, as he wrote to Henry, 'Bernard has studiously avoided understanding the various theories of the nature of dominance, and in the present book his own hypothesis does not stand up to close scrutiny ...' To Bernard he explained: 'Nowhere in the book do you produce evidence that dominance has ever evolved, with the exception of the situation in *Biston betularia*. Secondly, you do not discuss how modifiers selected for in the presence of the melanic can have been preserved at their same gene frequencies over long periods of time when the melanic was rare or absent ... However, your admirable observations on early melanics and your breeding results seem to contradict your thesis entirely. They cannot be explained solely by the accumulation of modifiers.'[147]

After an apoplectic phone call from Bernard, Henry talked Philip into softening his criticism, and Philip duly rewrote his paragraph about 'Bernard's contribution to science'. Kettlewell was the 'world authority' on melanism, he noted, and publication of his work was a major event. (*Nature* evidently agreed, featuring the review as its lead review.) 'If the author had taken a great deal more trouble with the book and the publishers had made sure the plates were inserted in the right places ... this book, instead of being a very interesting and useful volume, would have been a brilliant one,' Philip concluded, a judgment that failed to smooth the author's ruffled feathers. A few days later Bernard wrote to Philip, ostensibly to thank him, but quickly sank anew into a quagmire of grievances and self-pity.

I greatly resent your patronising remark ... 'if the author had taken a great deal more trouble ... etc.' Surely you realise it has taken me about ten years to write this book and in doing so it has more or less destroyed my soul. [A footnote here reads: 'By this I mean that as a direct consequence, it has led me into my own pathology.'] ... There was no lack of effort or 'trouble' on my part ... The ten years of writing also accounts for the references to earlier publications rather than the more recent ones. I wanted to get my work on melanism between two covers and without any shame I present this with all its deficiencies . . .[148]

A few more weeks of brooding only calcified Bernard's feelings of betrayal and isolation, and in January he lashed out at Philip again:

I agree with most of your criticisms in your Nature review but *NOT* the horrible way they are presented. No doubt now you have satisfactorily shot down the goose which laid the golden eggs . . . 'your own boys' will be able to make good capital from my 'speculative calculations which have never been published before' . . . Best of luck, but what a pity that most of them have minimal knowledge of the Lepidoptera in general.

In a meandering postscript Bernard took a pitiful parting shot at his old friend, who had recently been diagnosed with acute leukemia. 'My own view is that you would not have written such a review (in such a smug and bitter way) if you had been your usual self. I very much doubt the morality of someone like yourself undertaking such a project on the life's work of a friend.'[149] In general, reviews of *The Evolution of Melanism* ranged from poor to lukewarm. Bernard's torture was redoubled a few weeks later when a review by Jim Bishop, a former student of Sheppard's, let fly the demeaning phrase 'Kettlewell, until recently the leading authority on melanism in moths'. Bernard blamed Sheppard for poisoning the well; Philip protested that he had not spoken to Bishop at all.

What had troubled Ford and Sheppard in the rough-draft stage now troubled reviewers, namely the flimsiness and muddle-headedness of many of Bernard's central arguments.

He had made a mishmash of the evolution of dominance, everyone agreed, and whenever he wandered into evolutionary terrain he misspoke and exploded countless landmines. 'He wasn't an evolutionary geneticist,' explains Michael Majerus. 'This was a passionate man with huge, detailed knowledge of the Lepidoptera. There are vast amounts of data in there – the appendices alone are phenomenal – but any evolutionary geneticist who reads the book would find errors of emphasis on every fifth or sixth page, and would think, "This is a lepidopterist pretending to be something he's not."'

Miriam Rothschild is more straightforward in her criticism: 'It was a very bad book. He didn't cover up his lacunae.'

Some people had the sense that Henry had cursed the book in its cradle, like the bad fairy in Sleeping Beauty. 'It was Ford who ruined Kettlewell!' says his former field assistant Professor R.J. (Sam) Berry of University College London. 'He made him rewrite everything three or four times, by which time it had already been published by Ford. By the time Kettlewell's book came out, the whole thing was in the public domain. It didn't make much of a splash.' Bernard, for one, always believed Henry had sabotaged him. 'Father said it was spiteful and that Henry was jealous of his success,' said his son. 'He used to talk about it all the time. My mother got so sick of it all that she insisted he not talk about Oxford at the dinner table any more.'

Others viewed Henry more or less as he presented himself, as a long-suffering, patient benefactor. 'I never had any impression that Ford was jealous of Bernard's success,' David Lees says. 'He strived to give scientific rigour to Bernard's work through protracted discussion with him. Henry despaired about the results often. Bernard and Henry had a sort of love/hate relationship. Ford was an ascetic, an aesthete and a misogynist, but a real intellectual. Kettlewell was a larger-than-life *bon vivant* with a liking for female company, shooting and fishing, but no intellectual.'

It can never be known for certain to what extent E.B. Ford undermined Bernard, but given his temperament, his vanity, his taste for intrigue, and his need to pull all the strings, it is conceivable that he was not entirely sorry to see Bernard fall flat. Over the years he had accumulated many enemies, and he savoured these feuds as if they were fine, aged liqueurs; a controversy or a well-honed grudge always seemed to invigorate him. He had shown himself to be a cunning infighter who knew when to lie low and wait, sometimes for decades, for the right time to strike. When Bernard was embroiled in his dispute with Heslop Harrison, Henry had counselled him: 'Of course, they [Bernard's enemies] are hoping you will rise to the bait, and the thing that puts them in their place most is silence. When [Richard] Goldschmidt published a violent attack on me, I took a convenient opportunity of smashing him when I happened to be writing seven years later.'[150] Henry could employ language lethally like no one else, and even among the acid wits of All Souls it was said of him: 'His conversation was *dangerous*.'[151] Against Bernard, supposedly a friend and colleague, he would have known how to use a poison that left no trace.

Certainly it was true that Bernard's position at Oxford was, as he lamented to Philip, that of the goose that laid the golden eggs. Everybody, especially Henry, reaped enormous dividends from those eggs while consistently dismissing the goose. The peppered moths of Birmingham and Dorset were undeniably the jewels in the crown of the Oxford School of Ecological Genetics; Bernard's experiments were invariably listed first in Henry's reports to his bosses at Oxford, the Nuffield Foundation, and such. How often would Henry's hobbyhorses, *Maniola jurtina* or *Panaxia*, appear in a biology textbook? Virtually never. Even Sheppard and Cain's *Cepaea* took a back seat to Bernard's moths. 'How could this man be quoted in every single textbook in the world and not be respected?' says Rothschild. 'However you might criticize it, the experiment became famous.'

Bernard was an extraordinarily gifted field lepidopterist who was weak in genetics, statistics and evolutionary theory. How could he be expected to debate complex issues like the evolution of dominance with the kings of population genetics? It was Henry in the first place who pointed Bernard in the direction of the evolution of dominance, in which he became mired, and it was he who had first insisted that melanics had been around in pre-industrial times. Perhaps he did Bernard no favours by loading him down with his *idées fixes*. 'I think Ford had been speculating along very well defined tracks because he had in his mind that things ought to be this way,' speculates Laurence Cook. 'Kettlewell came into this from the outside and was groping, hoping that he had got the right end of the stick. In some cases he might have tried to produce the right sort of evidence and ended up producing evidence that was contradictory because he hadn't got it quite worked out in his mind.' In another place and time, a mentor less imperious and narcissistic than Henry might have found a way to give Bernard proper guidance, but Oxford graduate students in general were given free rein, and Henry had never been a hand-holder.

Henry had troubles of his own. He had suffered a mild heart attack in 1971, recovering quickly. More threatening were the barbarians massing right outside the gates now. It was taking more and more of his energy to fight back with letters to editors, and letters to influential friends asking them to write letters to the editor. While the textbooks were filling with happy tales of the peppered moth, evolutionary theory took a body blow from the Austrian-born logical positivist philosopher Karl Raimund Popper (1902–94), whose distinctions between science and pseudo-science powerfully shaped the standard criteria of the scientific method and who cited the theory of evolution as the most dismal example of nonscience masquerading as science. He viewed with particular scorn the fact that evolutionists had the gall to speak of a search for general 'laws'.

There could be no law of evolution, in the sense of a law of gravitation or Maxwell's equations governing electromagnetism, for it was a historical science, he decreed. Every event was unique and non-generalizable. Natural selection was 'metaphysical' and tautological, boiling down to the statement 'The survivors survive.' It violated the Popperian edict that proper scientific hypothesis must be falsifiable; one must be able to conceive an observation or test that would disprove it. By 1980 Popper would change his mind and acknowledge that evolution did have 'scientific character and could generate hypotheses that could be tested'. But his critique had already drawn blood, and some of his arguments found their way into a deft attack on E.B. Ford's empire, in the form of a 1972 review in *Nature*. Henry's nemesis Richard Lewontin of Harvard reviewed *Ecological Genetics and Evolution*, a *festschrift* book edited by E.R. (Robert) Creed, one of the acolytes of the School of Ecological Genetics, in honour of E.B. Ford. It was chiefly a collection of papers from the Oxford School's sphere of influence: A.J. Cain on *Cepaea*; E.R. Creed on melanism in ladybirds; Kennedy McWhirter and E.R. Creed on spot placement in *Maniola jurtina*; John Turner on Mullerian mimicry in burnet moths and heliconid butterflies; Bernard Kettlewell, C.J. Cadbury and David Lees on 'Recessive melanism in the moth *Lasiocampa quercus*', among others. At a lavish party at Miriam Rothschild's celebrating the book's publication Henry had basked in the attention like a sultan amid his favourite concubines, although he violently disliked his portrait on the frontispiece, painted by an artist commissioned by Rothschild. 'The whole thing is most splendid except for the ghastly portrait . . .' he wrote to Jim Murray. 'It may be art . . . [but] I do not have such narrow and sloping shoulders.'[152] To Murray, and everyone else, it seemed 'a very good likeness' – the bald pate, the beetle brows, the grimacelike closed smile, and, yes, the sloping shoulders.

In his review, Lewontin rejected Popper's first objection, that evolutionary theory was incapable of generating laws, but took up his charge that natural selection is unfalsifiable: 'Can one really imagine observations about nature that would disprove natural selection as a cause of difference in bill size?' he asked. '. . . Natural selection explains nothing because it explains everything.' Natural selection might be true, he argued, but how to verify it?

So if we chose 100 examples of variation between organisms . . . and tried to pin down the environmental circumstances responsible, we might succeed, after immense effort, in producing only two reasonably convincing cases. Then we would be in a Popperian pickle because we could not know whether natural selection was in fact a rare cause of evolution or whether it was a common cause but damned hard to demonstrate. Everyone would have wasted his time . . . It has been their hope [of many evolutionists] to explain so many cases of spatial and temporal variation . . . as to convince themselves and everyone else that natural selection is the chief agent in determining evolutionary change. In very large part this has been a British pastime, traceable to the fascination with birds and gardens, butterflies and snails that was characteristic of the prewar upper middle class from which so many British scientists came. E.B. Ford, the social and scientific quintessence of that tradition, was one of the earliest to devote his attention to the demonstration of natural selection.[153]

E.B. Ford's copy of this review, preserved in the archives of the Bodleian Library, is underlined in places, and in the margins the tiny creeping vines of his pencilled notes betray his agitation. 'He is evid. hoping to cast discred on the ext analysis of evolu by making out that evol. cannot be studied sci.'[154] The crack about the upper-middle-class British pastime and Ford's social class inspired a furious Fordian marginal snarl about 'presumptuous personal remarks about myself . . . he can have no [way of] knowing whether my ancestors were aristo, middle-class or working class.' (This was, naturally, a sensitive point for Henry.) Later, in what were likely notes for a rebuttal, he jotted: '. . . imply does not know enough science to write his review . . .'

The fact that Bernard's moths were singled out as one of the few meaningful Oxford School polymorphisms surely would not have pacified Henry. Making a distinction between two different kinds of selection, Lewontin cited Kettlewell's experiments as an exemplar of the good kind, 'functional selection', in which the character itself (melanic coloration, in this case) is of advantage to the organism in its relation to the environment. In 'tautological selection', in contrast, the observable character is accidental, and some unknown physiological difference accounts for the variation. This type of selection was nearly worthless in Lewontin's eyes. Any case of industrial melanism *not* ascribed to protection from visual detection – such as observations that 'melanic genotypes in heterozygotes, or in larvae, where the coloration is not an issue, simply leave more offspring' – is tautological selection, Lewontin wrote. Of course, Ford's primary contribution to industrial melanism was of this sort. The review continued:

If the spots on ladybirds [Robert Creed's line of research] are black in industrial areas and red in unpolluted areas, not because of the selective advantage of the colour itself, but because the enzyme that works better in polluted areas happens to give black pigment, then only in a formal and rather trivial sense is spot colour evolving because of natural selection. It will be a hollow triumph indeed for those who see natural selection as all powerful if it turns out that a very large proportion of morphological and behavioural evolution is nothing but the accidental concomitant of selection of polypeptides with greater heat stability.

If this were not insulting enough, from Henry's point of view, Lewontin's own book, *The Genetic Basis of Evolutionary Change*,[155] two years later would resume the battle and portray the Oxford School crowd as silly toffs with butterfly nets. (In fact, quite a few of the bright young graduates of the Oxford School of Ecological Genetics came from backgrounds that were far from privileged.) '. . . the British school,' he wrote, 'deriving in no little part from E.B. Ford, carries on the genteel upper-middle class tradition of fascination with snails and butterflies. These workers see the world as Darwin saw it, rich in diversity and, as confirmed Darwinians, they have assumed that there must be immense genetic variation available for adaptation through natural selection.'

The snipings of Marxist population geneticists from Harvard did nothing to dampen the British fascination with peppered moths. Jim Bishop, of the University of Liverpool, author of the notorious 'until recently the leading authority on melanism' phrase, was sampling frequencies of the different genotypes along a 54-kilometre transect from Liverpool to rural north Wales. He glued dead moths to trees at seven sites along the corridor to estimate predation rates. He also collected data on female fecundity and egg-laying, and did experiments to gauge the distance male peppered moths could fly (about two kilometres a night, he discovered). He fed all these data into a computer model simulating the rise, increase and spread of the *carbonaria* morph along this transect, starting with the assumption that at some time in the early nineteenth century the frequency of *carbonaria* was close to the mutation rate. (He also factored in a certain heterozygote advantage because everyone assumed the heterozygote must be fitter than either homozygote – an assumption that turned out to be wrong – and a small negative frequency dependent selection component, assuming that as a form became rarer, at either end of the cline, birds would search for it less.)

He cycled his model for 160 generations, or 160 years, as the peppered moth conveniently has one generation per year. He compared the genotype frequencies predicted by the model with the frequencies in wild samples caught at various points along the transect. The fit was far from perfect. *Carbonaria* was at a disadvantage (eaten more often than typicals) in an area where its frequency was well over 50 per cent. He concluded, in his 1972 paper, that 'additional selective forces need to be invoked' to explain the situation.[156]

A greater puzzlement was the high frequency of *carbonaria* in rural East Anglia, where the lichened trees were undisturbed by pollution, their trunks unsooted. (Bernard had proposed that the prevailing southwesterly winds were blowing industrial smoke in that direction.) In 1975, Robert Creed and David Lees, formerly Bernard's assistant and now at University College of South Wales, in Cardiff, conducted a predation experiment in East Anglia (with dead, glued moths) as well as one in the wood previously used by Bernard near Birmingham. In Birmingham, the birds got so adept at finding and gobbling up the moths that the experiment had to be relocated. (It is well known that birds can become entrained on prey, and some would say that Bernard, too, had created a bird feeder in the woods.) When the figures were sorted according to climate, there was evidence for selective predation only on the two 'wet' days. This was far from a confirmation of Bernard's results. In the predation experiment in East Anglia, the typical moths had superior survival on the light, lichened tree trunks, as expected, but this did not jibe with the fact that *carbonaria* comprised more than 60 per cent of the population. The birds were eating more of the *carbonaria*, but they were thriving anyway! Fitness and frequency were telling two different stories. 'We conclude therefore,' wrote the authors, 'that either the predation experiments and tests of conspicuousness to humans are misleading, or some factor or factors in addition to selective predation are responsible for maintaining high melanic frequencies . . . selective factors other than camouflage are important.'[157]

There was an evident gap between the predictions founded on Kettlewell's model, based on bird predation and crypsis, and actual observations. 'The discrepancy may indicate,' Jim Bishop and Laurence Cook reflected in a 1975 *Scientific American* article,[158] 'we are not correctly assessing the true nature of the resting sites of living moths when we are conducting experiments with dead ones. Alternatively, the assumption that natural selection is entirely due to selective predation by birds may be mistaken.' Even then some people knew that the stories and pictures in the textbooks were not quite true, yet it would be another two decades or more before the icon was seriously challenged.

It is an odd and disorienting experience to spend three weeks reading the letters of a stranger. It was a mild May in Oxford; the lilacs were in bloom everywhere, as giddy graduates in black robes with ermine cowls posed for photographs with parents and friends in front of the Sheldonian Theatre. Across the street in the Special Collections and Western Manuscripts Room of the New Bodleian Library, I was breathing in the bottled air of 1942, 1953 or 1966. (To obtain my Reader's card I had been ushered into a hushed office to recite aloud an ancient oath, swearing not to desecrate the library or 'kindle any flame therein'. I was told by friends that the oath used to be uttered in Latin.) The letters of H.B.D. Kettlewell, arranged alphabetically by the names of the correspondents, seemed fragile as moth wings within their mould-resistant boxes. Untying the satin ribbons binding a sheaf of letters between Kettlewell and a certain correspondent – Cyril Clarke, say, or E.B. Ford – I took leave of the present (the calming susurration of scholarly laptops) and immersed myself in a temporal stream that flowed ineluctably from the ebullient hopes of 1937 or 1942 to the sad setbacks of the Seventies. The next stack would unfold a different series of freeze-frames of the same journey, and I would traverse it in a single day, from the balmy high noon of a man's life to its tragic finale. After a while, I knew what was coming: all the bad months, the bad years, and it was eerie to contemplate the innocence of the letter-writer who did not have this foresight. Bernard Kettlewell's early letters were so charming, funny and irrepressible that it made the later ones all the more heartbreaking, and I often found myself wishing that it would all come out better, but of course it could not.

After the book he had laboured over for years emerged to barely tepid reviews, the exuberance seemed to drain out of Bernard. He officially retired in 1974, receiving a grant from the Royal Society to continue his work, and became an emeritus fellow of Wolfson College. (Retirement at the Oxford School of Genetics was often a formality, for the retired Henry was just as industrious as ever.) In the next year Hazel had a heart attack, and although she recovered, she was not her old self for a time, which was a problem, as Bernard required a great deal of looking after, as much for his inner torments as for his mounting physical infirmities. In June 1975, he did receive his DSci (Oxon) – a real Oxford doctorate – becoming at long last the right sort of doctor. Philip, who had slipped out of his remission from leukemia and was back in hospital, wrote to congratulate him, Henry gave him an obligatory little party at All Souls, and Bernard wrote some pleasant, jocular notes to his friends, but disappointment and sadness engulfed him.

June brought a happy family event, David's marriage. November brought a tragic one, Dawn's suicide. Since her divorce her compulsions and obsessions had worsened; she was obsessed with animals and would not use her oven because kittens were nesting in it. Now 'Darling Daughter', Bernard's nickname for his tortured elder child, was dead at age thirty-two, leaving behind a beleaguered ex-husband and two young children.

As for his career, Bernard's dashed hopes were embodied by one failure in particular. Although the Soviet Union had awarded him its Darwin medal in 1960, and Czechoslovakia its Mendel medal in 1965, these honours fell far short of the one he coveted, which was to have those magical three letters – FRS, Fellow of the Royal Society – after his name. Ford was an FRS, of course, and Philip Sheppard was, and there were at least five strutting around in the zoology department alone. The procedure for nominating someone as a Fellow is every bit as Byzantine and quirky as one might expect. Julian Huxley, who was a great friend and booster of Bernard's, had offered to sponsor him. Henry, however, had cut him off at the pass, saying: 'I'll do it.'[159] He dutifully nominated Bernard for the Royal Society three times, filling out the paperwork, gathering signatures, and putting Bernard's reprints in a box and delivering it to the Society. All the while, he confided to others that he didn't think Bernard had much of a chance.

Bernard was put up for the Royal Society for the third and final time in 1976. 'I have taken all Bernard's papers and books up to the Royal Society myself,' Henry wrote to Philip, who was in hospital again, battling septicemia, a complication of his leukemia. 'He very much wanted another shot at getting in, so he ought to be given the chance, though I don't think he will make it. It seems to me that he has missed his market: largely by taking up all the time to write that negligible book about Darwin with Julian Huxley.'[160] Henry had told Arthur Cain much the same thing, that Cain would never be accepted into the Royal Society because he had taken too long to write a book, but Cain had become an FRS in spite of his prognostications.[161] Bernard did not, and this time he knew there would be no more chances. He was a beaten man, and Oxford became a bitter topic that Hazel and Nanny tried to ban from the dinner table.

Though Henry professed himself concerned for Bernard and seemed to be doing his best, a more cynical reading of his behaviour suggests itself. Bernard himself came to believe that Henry was secretly sabotaging him – 'building him up and then cutting off his legs', as his son David put it – and there is no doubt that Henry was astute enough to imply, without actually saying so, that Bernard was not quite up to the mark. Although there is no way to know exactly what happened behind the scenes, in the Bodleian archives there is a letter from Bernard to Julian Huxley in 1965 referring to 'the private remark you made to me in connection with the

Royal Society', that seems to contain a clue. Here is what an obviously agitated Bernard wrote:

I must say I was somewhat shattered as the facts are entirely different from those quoted to you. While Henry has always been of the very greatest help to me, he would be the first to admit that, apart from breeding small numbers of typical and melanic forms in the laboratory (in which he showed that the black forms were hardier), he has done absolutely no work on industrial melanism. The designs of experiments, all the large-scale field releases, the survey of frequencies of all British species which manifest industrial melanism, were of course all my own idea and efforts ... [This idea] has probably arisen from Henry's latest book, 'Ecological Genetics', in which he devotes more or less a chapter to each of his colleagues ...[162]

It is not too great a leap to infer that what Huxley had confided was that people were saying that the real force behind the peppered moth work was Henry, not Bernard, a suggestion that by all evidence was untrue. Bernard may have been merely preserving the forms of politeness by saying that Henry 'would be the first to admit' this, for it must have occurred to him that it was most likely Henry himself who had leaked this rumour to the Royal Society.

'Not getting into the Royal Society finished him off, I think,' his son, David, told me. 'He was honoured by the Russians; he was honoured in America and Italy and all over the world, but at home he felt totally unappreciated. All he wanted from Oxford was a thank you, some acknowledgement of all his work. In the last five years of his life he gave up.'

Philip Sheppard finally lost his valiant three-year battle with leukemia in 1976; ever stoical, he had been dictating papers on swallowtail genetics from his hospital bed on the morning he died. Robert Creed died very suddenly of cancer of the thymus in 1978. Bernard was terribly saddened by both deaths, and by this time his own ailments were restricting his life more and more. He had injured his back years before when, climbing a stepladder to reach something in a medicine cabinet, he had fallen backwards onto the bathtub taps. Then he repeatedly reinjured his spine in the field. 'He was always carrying these very heavy generators around,' David recalled. 'In Scotland we would have to drag them up hills. He would climb over fences with them if that was the only way to get them there.' His back had become a constant torture; he was hospitalized several times, invariably becoming the victim of multiple iatrogenic complications. He was desperately tired now, and his fatigue and melancholy all but eclipsed his fun-loving side.

'The only time he was happy was in the field,' says Rothschild. 'He was such a good naturalist. I went on collecting trips with him, and no one could touch him. He had the golden touch. He bred things and his things hatched out; mine didn't. It's a peculiar kind of intuition, like a gardener's green thumb. Otherwise, he was a very unhappy man. He was unhappy because of his temperament.'

His lifelong pattern of careening from crest to trough and back again suggests a cyclothymic or bipolar mood disorder, a familial curse, perhaps, that was visited on his unfortunate daughter in an even more lethal form. It is equally possible that his difficult temperament had its roots in the emotional tundra of his childhood, which seems to have stifled him in some way and instilled a strain of ineradicable self-doubt. 'He was dominated a hundred per cent by his mother,' said David Kettlewell. 'I think this affected him very much. He never had any family love, never had any compassion or friendship from his parents. He always felt a bit unloved.' This, in turn, impaired his ability to be a loving parent to his children. 'In truth, he wasn't a very good father,' David continued. 'He tried, but because he had never had any love, he didn't know how to give it. In front of other people he was usually warm, but when the doors closed he was a different person. I think he was too involved in his work. He was very strict; he expected more of me than I could fulfil.'

In July 1976, Bernard confided to Cyril Clarke: 'I am still slowly but steadily going downhill . . . becoming more and more weary and more and more feeble. I could willingly spend a large part of the day sleeping.' In January 1977 he was in the Radcliffe Infirmary again. 'I had no fewer than 13 different tablets last night,' he wrote Miriam Rothschild, 'and was woken up at 6 am for an investigation of the last remaining orifices for evidence of what I have or haven't got.' A month later, after breaking two lumbar vertebrae in a fall, he was laid up once more and given great quantities of 'horse pills', which he thought turned him into a 'zombie' for months. Feeling 'depersonalized', he worried that he was losing his mind, that he was developing Alzheimer's disease.[163] His back pain became unbearable. In the summer of 1977 he spent several weeks in the hospital before coming home to a restricted life with Hazel and Nanny.

His favourite dog was ailing, as well, and a vet gave Bernard pills to put her down, but instead of producing a quick euthanasia the medication caused the beloved pet to go horribly berserk. Stricken with horror and remorse, Bernard could not erase the image from his mind. 'He couldn't stop talking about it. He was obsessed for weeks,' Miriam Rothschild recalls. Bernard had always had unreasonable fears and phobias, but now they were no longer confined to the back rooms of his consciousness. On stormy nights he would lie awake worrying about stray dogs that might be out in the cold and dark. His mind had become a hell-realm. He could still transcend his troubles now and then in the company of his beloved insects, but in July 1978, while collecting, he fell out of a birch tree and broke his back again, a blow from which he would never recover.

His letters from this period are painful to read. In February 1979, he wrote to Cyril Clarke: 'I feel that in fairness to Hazel, I cannot go on as I am at present and it is a somewhat miserable life.'[164] He was placed on a waiting list for surgery at St Bart's in London and told the delay could be as long as two months. Around this time, he received a letter from the British Museum of Natural History informing him that his cherished Rothschild-Cockayne-Kettlewell collection, now known as the National Collection, was being dismantled, and all the collections of Lepidoptera amalgamated. Thus the other part of Bernard's legacy was annulled at one blow. 'Cyril Clarke wrote a very nice letter,' David acknowledges, 'but my father had given up by then. He was suing enemies round the corner.' When David left England to take up a job in Connecticut, Bernard told his son bluntly: 'You won't see me again.'

He died on 11 May 1979, his entry in *The Dictionary of Scientific Biography* states, 'apparently from an accidental overdose of the painkiller he was using to relieve a back injury sustained during fieldwork'.[165] The overdose was not an accident. Bernard was a doctor and knew exactly how to commit suicide efficiently. Whether the precipitating factor was intractable back pain or dread of Alzheimer's disease or general discouragement or all of the above, he did what he had always matter-of-factly told his friends he would do when all hope was lost. Someone who had seen him that day found him cheerful, even ebullient, and found it hard to believe that he could have taken his life by the end of the day. But his old friend Cyril Clarke said that Bernard's mood swings were so extreme that it was highly possible for him to have been on top of the world one moment and suicidal the next.[166] Among the flood of loving obituaries that appeared following his death was one by Clarke in the *Entomologist's Record*, which had been the showcase for many of Bernard's entomological triumphs:

Bernard in some ways belonged to the last century, when field work by first-class amateur naturalists, some of them like him, truants from medicine, built up the taxonomy of the lepidoptera. He it was, however, who showed by his genetic studies how right these amateurs were. Bernard also belonged to this earlier generation because then people were not afraid to enter into full-blooded controversy (his outspoken answer to J.W. Heslop Harrison on melanism is an example of this) but there was never anything underhand or scheming in his attacks, he just said what he thought. In character he was touchy, argumentative and often maddening, but he could laugh at himself and was extremely good company. He had one particular edge on us all – everyone loved him.[167]

Hazel had for some time wanted to move out of the sprawling house at Steeple Barton, but Bernard had been too infirm and too inflexible in his last years to effect such a big change. After his death, Hazel and Nanny moved together to the Sussex coast, where Hazel lived until 1993 and Nanny to the advanced age of 106. In the years after Bernard's death Hazel wrote to Miriam Rothschild that on harsh and stormy nights she would lie awake and worry about Bernard being out in the cold, just as he had lain awake worrying about lost dogs.

It is not reported how Henry took Bernard's death, except that he pronounced him a 'coward' for taking his own life.[168] For several years he had been fed up with Bernard's infirmities and physical decline, which he viewed as slackness and lack of character. Henry himself soldiered on through mild heart attacks, colon surgery and hip replacements, arriving at his office sharply at nine o'clock every morning, and going off in wellies and a brown trilby hat to catch butterflies in season in Cothill and on the Devon–Cornwall border. Nothing seemed to deter him from being Henry, an ageless late-Victorian tintype whose attitudes and beliefs, like his looks, were curiously impervious to change. Henry had looked middle-aged when he was young, and, strangely, he still did.

Yet the world that Ford built was under attack from many quarters. Toward the end of the 1970s the monolithic selectionist worldview that had dominated evolutionary biology began to erode a bit. In 1978 Henry's old *bête noire* Richard Lewontin was invited to speak at a Royal Society symposium on adaptation, but Lewontin hated to fly and was too busy to travel to London anyway, so it was decided that he would write a joint paper with his Harvard colleague, the paleontologist Stephen Jay Gould. Gould ended up writing most of the paper, called 'The Spandrels of San Marco and the Panglossian Paradigm: A Critique of the Adaptationist Program',[169] and flew to London to present it. He was already a maverick, having proposed, with Niles Eldredge, the theory of punctuated equilibrium in 1972. According to this theory, evolutionary progress is not gradual and smooth; rather, most species are static for long periods, and change, when it occurs, usually occurs rapidly, concentrated in small population groups. (Punctuated equilibrium can be recognized as a latter-day form of saltationism, the school of jumps.) Although trained as a typical adaptationist, Gould had developed a cynicism about what he would dub 'naïve adaptationist Just So stories', the practice of demanding an adaptationist explanation for everything. His Spandrels paper asserted that 'an adaptationist programme has dominated evolutionary thought in England and the United States during the past forty years. It is based on faith in the power of natural selection as an optimizing agent. It proceeds by breaking an organism into unitary "traits" and proposing an adaptationist story for each considered separately . . .' While Gould did not deny the power of natural selection, he took issue with the 'near omnipotence of natural selection in forging organic design and fashioning the best among possible worlds' – a view he traced to A.R. Wallace but not to Darwin. The symposium's moderator, Arthur Cain, who was a veteran of the Oxford School of Ecological Genetics and as adaptationist as they come, was observed fuming, and subsequently devoted his summary of the programme to a counterattack on this paper and its authors.[170] It is not known whether Henry attended the conference, but he certainly would have shared Cain's outrage. (This particular paper did not explicitly make fun of the beetle-collecting habits of upper-class Englishmen, but in subsequent writings Gould managed to get in a few digs at the Oxford school's obsessions.)

Gould would go on to insist that during the 'hardening of the synthesis', as he called it, other evolutionary processes had been steadily overlooked, including biological constraints (developmental constraints, for example), nonadaptive processes, rapid speciation, and separate selection at the level of species. Gould, his collaborator Niles Eldredge and others were beginning to question whether all the meticulous observations made by Ford *et al.* – the tabulation of wing spots and different alleles – could reveal much about the big picture. Even if natural selection operated within populations, turning white moths into black moths, for example, was this a unitary, universal mechanism that could account for the success of one species or genus or family over another? The fact that most major extinctions were the result of natural disasters, which could hardly have respected one organism's refined feelers or another's warning coloration, was an awkward fact for extreme adaptationists. At the end of the Permian Era 225 million years ago, a doomsday cataclysm (the greatest of five great extinctions on Earth) wiped out an estimated 96 per cent of all marine life and three quarters of vertebrates on land, and the lucky survivors became the ancestors of us all. Yet the survival of this tiny band and the extinction of all the others had nothing to do with 'fitness' or 'adaptation'. It was blind chance.

By this time there were beginning to be two camps. The strong selectionists, Darwinian hardliners and genic reductionists included the E.B. Ford school, Richard Dawkins and Maynard Smith in England, and George C. Williams in the US, while the other camp included the likes of Lewontin, Gould, Niles Eldredge, as well as the neutral molecular crowd. It was never the case that the anti-hardliners were anti-evolutionist, however much creationists may have misunderstood their views.

While the other fathers of the modern evolutionary theory were wont to pontificate on lofty matters such as the rise and fall of species, man's place in the universe and the meaning of life, E.B. Ford remained a resolute miniaturist, wholly uninterested in the big picture. When he did mention speciation, it was only cursorily, and for the most part he seemed to find the subject, like metaphysics, unworthy of serious interest. 'The title of Darwin's book . . . is in fact an unfortunate one,' he writes in *Taking Genetics into the Countryside*, his last book. 'It leads sometimes to the idea that evolution consists only of speciation. Whereas adaptation, in colour-pattern, habits, or physiology; variation or stability [and a dozen other things, which he listed] are all aspects of evolution.'[171]

Henry was content within his own untroubled sphere, where order reigned and all the rules were known. By the late 1970s all sorts of changes were invading Oxford, however, and the comforting celestial order was breaking down even at All Souls. The Franks Commission, charged with modernizing Oxford in the mid-1960s, had issued a report claiming that All Souls was sitting on a fat fortune and contributing next to nothing to the communal life of the university, and that its practices were characterized by a Macbethian 'infirmity of purpose'. Eventually, John Sparrow was forced to resign as Warden, and slid into flamboyant dissolution and alcoholism.

By the 1970s there were rumblings about the admission of women, an idea that filled Henry with the same depth of horror that Aristotle might have felt about female philosophers. He sat down in 1978 and penned a letter to the Warden and Fellows on 'The Abilities of the Sexes, and the Admission of Women to All Souls', in which he argued that there had been no great women composers, painters, poets or air traffic controllers. 'Such distinctions apply widely, far beyond the confines of art. For instance, ability to reach a quick and reasoned choice. This can be found at an Airport. On the one hand in the capacity of women to track approaching planes; on the other, men are indispensable for deciding the correct runway to be used by landing aircraft.' Henry had evidently given the matter a great deal of thought. From the standpoint of genetics, he insisted, 'the human sexes differ as much as members of separate species because of their chromosome outfit that maintains the dissimilar balance of their sex-genes . . . Women pass out of the maximum learning period, as they enter it, earlier than men. It is far from true that outstanding women undergraduates, compared with men, make brilliant researchers.'[172] The fact that Henry sincerely believed this did not make it any less galling to many people, including Miriam Rothschild, who had been responsible for his admission to All Souls, where he was known to the staff as 'Mrs Lane's fellow'.

It struck her as a 'very vituperative letter. People said to me, "You have backed a very funny person." I wrote them a letter about this. I always pretended to be a retired army colonel, Colonel Jackson, and always signed my letters "God save the Queen".' Despite Henry's campaign, the dreaded 'female women' were admitted to All Souls, first to the dining hall and then as fellows. Nobody alive knows exactly how Henry had conceived an alienation from women so extreme that they seemed to him to belong to a different species. His friends noticed that, while he invoked his father with fondness, dedicating *Butterflies* to him, he never spoke of his mother. 'He never spoke about his childhood, never mentioned his mother,' says Rothschild. 'My own view was always that the twisted part of Ford's personality had something to do with his mother.'

In his waning years, Henry was befriended by John Haywood, formerly the photographer at the zoology department, and they produced a book together, *Church Treasures in the Oxford District*, in 1984. Though well reviewed by Henry's friend A.L. Rowse in the *Spectator*, the *Oxford Times* accused him of many inaccuracies.[173] Henry continued his lifelong study of fogous, now in the Algarve in Portugal. In 1985, a year after his first hip replacement, his housekeeper died. Having dismissed a successor whose habits interfered with his, Henry lived alone, depending on the kindness of a few friends and relatives. The same year, a burglar broke into 5 Apsley Road and stole his heirlooms, which he listed 'as two Ruskins, a little sketch by Constable, a Gainsborough, and a nineteenth-century Chinese fan'. His attachment to these treasures had been extreme – Miriam Rothschild recalls that 'he was constantly having me to lunch, and he would go around showing me his treasures, his pictures, and his jewels. He was terribly proud of his possessions' – and their loss must have been as traumatic as an amputation. During his last visit with him, at All Souls, Ernst Mayr found him lonely, sad, and feeling unappreciated. Yet in spite of glaucoma, severe asthma and a heart condition, Henry never stopped working. He worked for two years on a book on the causes of extinction, envisioned a new *Church Treasures*, and persisted in the usual jurting and *Panaxia* excursions. 'The last time we went jurting he collapsed in the bathroom,' says Haywood. 'Bunny [Dowdeswell] and I had to help him to bed.' On Thursday, 21 January 1988, according to Sir Patrick Reilly, 'he dined in college as usual, but was, I am told, unusually silent. He died in his room soon afterwards. He could not have had a better death.'[174] He was eighty-six.

Who was the real Henry Ford? One person who knew him suspected that he 'parcelled himself out to people – I don't think anyone had the whole story'. So difficult was it to penetrate the multiple layers of Henry self-mythology, to sort out facts from legend, that it took Bryan Clarke years to research and write Henry's biographical memoir for the Royal Society. The nimbus of mystery around him suggests that Henry was guarding a secret, or several, and perhaps it would have taken a private detective to get to the bottom of it.

'He did not reveal the school he went to, probably because he felt it wasn't grand enough,' Clarke remarks, noting that some of his friends thought he'd gone to Malvern, while others associated him with other public schools. 'I found out he'd been at St Bee's in Cumberland and talked to the headmaster of the school. He found no record of him except that he'd come and gone.' He may have kept as low a profile as possible, for a boy as peculiar as Henry was almost certain to have been tormented; he would not have been good at games, and his high-pitched speaking voice, effeminate mannerisms, and nerdy, solitary habits would likely have marked him as a pariah.

He was always chary of personal details and recoiled at the very notion of biography. When Joan Fisher Box, 'one of Ronald Fisher's many female daughters', persuaded him to reminisce into a tape recorder about her father, he complied grudgingly. 'Nor am I quite happy about biography,' he complained to Philip Sheppard. 'It is in his work and thoughts, rather than in himself, that the true interest lies.'[175] His squeamishness at the thought of biographers snooping around in his life or artifacts may have been due in part to his sexual secrets, but probably not entirely. Homosexuality was hardly a novelty among the Oxford intelligentsia; Henry's All Souls friends John Sparrow and Leslie Rowse, among others, were not at all shy about their homosexual lifestyles, and Henry had only a toe in the closet, after all. In 1962 Miriam Rothschild was gathering signatures for a petition to consider decriminalizing homosexuality in Britain. 'When I approached him he was very brave. He never hesitated. "You're quite right," he said. He got Ronald Fisher and Julian Huxley to sign it. Ronald Fisher said to Ford, "I will sign it because Jesus Christ would have signed it." Henry imitated the way he said it.'

Prior to a 1974 celebration of the Synthesis, the conference organizers, Ernst Mayr and William Provine, mailed questionnaires to all the participants, including E.B. Ford. Under 'personal data', Henry wrote touchily: 'I find here at least one question which does not appear relevant: i.e., the *place* of my birth (!) [This he listed as "The Manor House, Papcastle, Cockermouth, Cumberland, England".] It is not so easy to see what this has to do with the matter at hand. At any rate, it is unimportant indeed compared with other personal items for which no space seems available.'[176] By this he evidently meant his medals and prizes, which he proceeded to list, starting with his FRS. Something in the phrasing of his invocations of The Manor House in Papcastle suggests that, rather than owning it, his family was attached to the house in a more peripheral capacity. 'I wonder if his father was the vicar there,' says Bryan Clarke.

In answer to question 12, 'Other personal data that might be of relevance to your contributions to the synthesis', Henry wrote: 'I was Attaché at an Embassy before coming to Oxford. Until then, I was a Classical Scholar who had worked on evolution and archeology (my home was on the views of a Roman fort) on my own. As a boy I travelled repeatedly in Italy and Sicily.' Thus, by his own account, the teenage E.B. Ford was a classical scholar, evolutionist, archeologist, and an attaché at an embassy. 'That's interesting,' mused Bryan Clarke, when I brought this to his attention. 'I can't imagine how a seventeen-year-old would have been an attaché at an embassy.' Many of Henry's acquaintances were under the impression that he had been in the diplomatic service at some time, but Clarke's inquiries indicated he had not. Because of Henry's many cryptic comments about his war years – when Jim Murray once asked him why he had not gone to the Scillies during wartime, he answered: 'Some of us were needed *at once!*'[177] – many people believed he had done intelligence work, perhaps as a codebreaker at Bletchley Park. Miriam Rothschild, who was at Bletchley, states with assurance that Henry was certainly not. Furthermore, 'there are eyewitness accounts of wartime tutorials in Oxford, during which he knitted woollen balaclavas and socks for the Navy,'[178] according to Clarke, and his *Panaxia* fieldwork and publications continued uninterrupted throughout the war years.

He left strict instructions for all of his papers and correspondence to be destroyed after his death. Given that E.B. Ford was one of the major figures of twentieth-century evolutionary science, a former departmental secretary could not bring herself to comply, and eventually his papers were turned over to several friends and colleagues to sort and purge. What survives as the E.B. Ford Collection in the Bodleian Library is a highly sanitized, rather dull set of documents, thoroughly disappointing to the biographer (though, fortunately, shards of Ford's personality can be found in the archives of other scientists).

Henry erased his real self well, but not perfectly. Although E.B. Ford was known as a stickler for accuracy, meticulous to a fault, there had been rumours circulating that the final counts of the spots on the *Maniola jurtina* butterflies sometimes did not match the scoring sheets that were found in the inn's waste basket. Then there were Henry's fabled breeding experiments from the 1940s, when he had crossed scarlet tiger moth phenotypes and supposedly demonstrated the evolution of dominance. The normal practice is to save the 'breeding material', the pinned specimens from each generation, as a record. 'One would expect that people would keep their material and give it to a museum,' Clarke tells me. 'He actually destroyed it.' Did Henry really accomplish a complete 'dominance breakdown', with a full range of intergradations, as he had claimed, or had he merely said he did?

And there was something else. The account in his Royal Society memoir of Ford's pathbreaking work on *Gammarus chevreuxi*, which launched his career, carries a footnote: 'These researches have a statistical interest that is more appropriate for discussion elsewhere.' In this work Ford had shown for the first time that genes control the time of onset and rate of development of processes within the body. He was looking at genes that controlled eye colour, and found that they segregated out in a proper Mendelian ratio. It was Ford's misfortune to have a mathematical biographer, for Clarke re-analysed Ford's statistics and found that 'the correspondence between his results and theoretical expectation is too close.' Ford got the results he wanted, *exactly* the results he wanted. This a red flag, not unlike the celebrated case of Gregor Mendel himself, whose pea plant data were too close to perfection. The person who made the 'abominable' discovery that Mendel's data had evidently been doctored – statistically, the deviations from expected ratios were too small – was none other than Ronald Fisher, who immediately wrote to his intimate friend, E.B. Ford, to confide the 'shocking experience'.

Scientists are extremely reluctant to accuse other scientists of 'fudging', the highest crime in science, and Clarke does not go so far as to accuse E.B. Ford of fraud. 'His results may have been conditioned by his expectations. If you have to score colours, which is what he was doing, it's easy to get to thinking, "That's yellow" and "That's red." It is an understandable failing, but it is a failing.' Haunted by misgivings about his discovery, Clarke has worked up a paper on the subject but still hasn't decided whether to publish it. 'I showed it to a few people. The men said, "It is your duty to publish it." Women said, "It's unfair to publish this when he can't defend himself."' He discussed the dilemma with some friends of Henry's at All Souls and then regretted having done so. 'They were going to have a plaque that said "E.B. Ford: Geneticist and Raconteur" and they changed their minds when they heard this. Much care is needed in talking about it. His conclusions are still valid. I wouldn't like to give the impression that it means the whole of ecological genetics is no good.'[179]

The word 'brilliant' is invariably used in connection with Henry. While still in his twenties he had conceived an audacious project: to gauge the power of natural selection by detecting its operation in nature. He invented a new science, ecological genetics, which, due in large part to the endeavours of some of his brightest acolytes, was successful in defining the way Darwinism was practised for several decades. His death, it was commonly agreed, signalled the end of an era. As a scientist, intellectual, and above all raconteur, his powers deeply impressed almost everyone, yet he suffered the common fate of brilliant talkers in being inadequately memorialized in print. It is also the case that he did not earn a first-class degree from Oxford; that his writings, while renowned for their clarity, are sometimes boring and repetitive, his opinions rigid; his lifetime genetics work, with *Panaxia* and *Maniola*, yielded ambiguous results. It may be that E.B. Ford's big secret was that he wasn't quite as smart as everyone thought he was.

At his request, E.B. Ford's ashes were scattered on a grassy Cotswold knoll, leaving him in the eternal company of the meadow butterflies that had been his life's work. The symbolic association between butterflies and the soul was well known to Henry. In his 1981 book *Taking Genetics into the Countryside*, he wrote an uncharacteristically meditative passage on the symbolism of the peppered moth's transformation:

One wonders what symbolism country-lovers would attach to that transformation if they observed it . . .

To the ancient Greeks, Ψυχή [psyche] stood for the human soul or vital spirit (animus) and also for butterfly. (I hardly think the Greeks distinguished moths from butterflies.) I remember a Greek altar [that] showed a caterpillar, no doubt denoting the earthly body; the chrysalis, that body in death; and the flying insect, the soul of the departed . . . And Thomas Hardy told me [how like Henry to drop a famous name!] of a Dorset legend that a white moth escapes from a Man's mouth at the moment of death: the symbolism of the soul and of the white insect associated so far apart in time and in space, more than 2,000 years and at opposite ends of Europe. So, too the change in the darkened countryside and the moths darkened there may well be taken, even by those with scientific knowledge, as symbolic also.

After Henry's death, the peppered moth would undergo yet

another transformation. Having served as a symbolic vessel of the highest aspirations of Ecological Genetics for thirty years, it would find itself struggling in an uncertain post-Fordian world, in the barbarous lands across the ocean.

PART III

II

'It was a bird feeder!'

Everyone loved the peppered moth. The ritually repeated refrains about *British moths . . . industrial soot . . . classic demonstration* cropped up in every mention of evolution and natural selection, so often that they were like an incantation, a skipping-rope rhyme, a fragment of a show tune that everyone knows. Through the textbooks the moths had become embedded in the collective consciousness. They were a teaching story, and you might as well try to eradicate $E = mc^2$ or Newton's apple or Galileo at the Leaning Tower of Pisa. Everyone from the most celebrated population biologists to textbook writers to high school biology teachers trotted it out whenever a striking example of evolution was needed. The quotes stacked up like raves on a musical marquee in Times Square, or the blurbs on the flyleaf of a bestseller:

- '. . . the clearest case in which a conspicuous evolutionary process has been actually observed.' – Sewall Wright, 1978.
- 'The most spectacular change ever witnessed and recorded by man . . .' – Philip Sheppard, 1958.
- 'We should expect to find the most rapid evolutionary changes in populations suddenly exposed to new conditions. It is therefore natural that one of the most striking changes which has been observed in a wild population . . . is the phenomenon of "industrial melanism".' – John Maynard Smith, 1966
- 'One of the most striking examples of observable evolution is the phenomenon known as industrial melanism.' – entry in the *Encyclopedia Britannica* written by Sir Gavin De Beer, 1974.

Twenty-three years after Bernard Kettlewell's suicide and thirteen years after E.B. Ford's ashes were scattered over an Oxfordshire meadow, introductory biology students all over the world are still routinely squinting at photographs of peppered moths and memorizing the name 'H.B.D. Kettlewell'. Thus it is shocking to hear someone say:

'You can tell by looking at these that they're fake.'

Ted Sargent, emeritus professor of biology at the University of Massachusetts, Amherst, is inspecting the photos of peppered moths in a textbook. 'These moths are not in a resting attitude, because the antennae are out,' he explains. 'When they rest they pull the antennae back under the forewings and their wings are back in a triangular shape, not spread out like this.'

With two private colleges and a state university in its midst, Amherst is probably denser in academics than Oxford, but, apart from the odd Derrida paperback in the pocket, professors and graduate students tend to blend in with the other soccer moms and dads. Traditions here are less entrenched by several centuries, and it would be unusual to find someone as donnish and otherworldly as E.B. Ford in the movie line at the multiplex

– which is a few miles down the road, because Amherst zoning laws do not allow multiplexes to mar its nineteenth-century atmosphere. The town's namesake, Lord Jeffery Amherst, hero of the French and Indian War, is an embarrassment to modern, politically correct Amherst, since one of his feats of diplomacy was to sell smallpox-infested blankets to Amerindians. Although his name adorns a town, a college, an inn and a bookstore, it seems that Lord Jeffery never actually set foot in Amherst, and if the town has a presiding deity it is Emily Dickinson, whose plain, spinsterish countenance looks out onto eternity from the walls of bookstores, libraries and coffee houses.

Sargent knows far more than the average biologist about Emily Dickinson's poetry, particularly its abundant bird imagery; he has tallied the mentions of robins, phoebes, hairy woodpeckers, whip-poor-wills, bobolinks, orioles and hummingbirds in her verses. For the most part, though, we talk about other birds, the birds that may or may not eat peppered moths: 'Those experiments in Birmingham and Dorset were just terrible! Kettlewell set the moths out in huge numbers, way above normal levels for those environments, and then acted surprised when the birds came and ate them. It was a bird feeder!'

Sargent is fit-looking at sixty-four, square-jawed, square-shouldered, with iron-grey hair and wintry blue eyes, yet there was something about him that made him fade into the wallpaper. Only the striking emerald-green cover of *The Evolution of Melanism* by H.B.D. Kettlewell resting on his table alerted me to his presence. When he fixes his eyes on the pictures, though, the vagueness is replaced by the cool, appraising gaze of a jeweller looking for flaws in a diamond, or a seasoned homicide detective scrutinizing crime scene photos. As far as Sargent is concerned, these photos of Bernard Kettlewell's moths might as well be crime scene photos.

'These are pinned specimens. They are dead; they were pinned up and then taken down. Sometimes they take a live one and let it crawl around and then take a picture of it. But they're all fake; no one has found one on a tree trunk. Who's going to find a moth out there like this, let alone two demonstrating crypsis?'

Sargent has been waiting a long time for someone to listen to him, and his words spill out as from someone just released from solitary confinement. What he has said about industrial melanism has shaken up a paradigm and made quite a few people uneasy. 'Industrial melanism is a club, and the fact that Ted wasn't in their club hurt him profoundly,' according to Lincoln Brower, the monarch butterfly expert, who is an old friend. Brower is yin to Sargent's yang, a charismatic figure on PBS nature specials, blue-ribbon scientific panels and Smithsonian covers, as much the golden insider as his friend is the overlooked outsider. 'They ignored him because he isn't British. The mindset is so strong in England. "We've got the paradigm." It was their loss. Ted has massive data. He's a very good scientist.'

If you were to say that Ted Sargent's career was derailed by the peppered moth, you would not be entirely wrong.

When Kettlewell carried out his legendary experiments, Theodore David Sargent was still in high school in Peabody, Massachusetts. Sargent's father had heard Alfred Russel Wallace lecture; his grandmother had heard Darwin speak; and Ted, like the young Bernard Kettlewell, was a boy with a moth collection. In the swamps, fields and vacant lots of his North Shore town, just up the road from historic witch-haunted Salem, he absorbed natural history as easily as modern boys master PlayStation codes, spending his summers at a succession of nature camps and Audubon camps, where libidos ran surprisingly hot. After graduating from the University of Massachusetts in Amherst in 1958, he earned a PhD in ornithology at the University of Wisconsin, and it was as a bird man that he was hired by his alma mater in 1963. By this time, people were starting to talk about Kettlewell's work, and Sargent was as impressed as anyone: 'You had the complete story. You had a gene which appeared in increasing frequency. You had a selective agent, the birds. The whole evolutionary picture was there. It was so powerful.'

Then a young, unmarried assistant professor living in an upstairs apartment, Sargent meanwhile fell under the spell of the moths in the backyard. Summers in the Connecticut River Valley, often as steamy as an orchid greenhouse, would come as a shock to the system of any visiting professor from drizzly Oxford, and even long-time residents find sleeping difficult on the hottest nights. For Sargent, whose focus was the moths of sultry midsummer, it was a great season for sugaring, which he did by dissolving two pounds of dark brown sugar in about six ounces of stale beer and painting the mixture on tree trunks. A cheap commercial elderberry wine worked well, too. His bait traps would attract over a dozen species of underwing moths (*Catocala*), the largest and most showy of the New England moths, and these 'seduced' him. He had met them before, at the age of twelve, when he borrowed a copy of W. J. Holland's *The Moth Book* from the Peabody library to savour the rows of moths displayed in the colour plates. Most moths in the Noctuid family, drab greyish or brownish, seemed lacklustre after the ornate silk moths and tiger moths he had contemplated – until he arrived at plate after plate of resplendent *Catocala*. He read what Holland, that gushing Victorian voluptuary of the natural world, had written about them: 'Did you ever reflect upon the fact that the wings of many moths, which lie concealed during the daytime, reveal their most glorious coloring only after dark, when they are upon the wing? ...

The fore wings are so colored as to cause them, when they are quietly resting upon the trunks of trees in the daytime, to look like bits of moss, or discolored patches upon the bark . . . one of the most beautiful illustrations of protective mimicry in the whole realm of nature.' But their hindwings: 'Oho my beauty!' Holland famously exclaimed in print. Banded with pink; or edged with crimson, yellow, orange; or snow-white on a background of jet-black; they were glorious, a pageant beauty disguised as a nun. 'Questions rushed into my mind,' Sargent recalled. 'Why were these moths so large? Why did they have such spectacular hindwings? And why were there so many kinds?'

There were seventy-one species of *Catocala* in the eastern United States alone, he learned, and almost no published literature on their life histories, behaviour or ecology. Here was a nearly virgin field where a newcomer might manage to get published. Soon he was jettisoning two years of bird research and dreaming up ingenious testing apparatus for moths. Nowhere near tenure, he decided to begin again at square one as a lepidopterist.

One day he started up a conversation with a slim, dark-haired girl named Kathy who was sketching birds in the corridor outside his office. They were married in 1965. A registered nurse, technical illustrator, and painter turned prize-winning quilter, Kathy Sargent gave birth to their son, David, in 1967, followed by daughter Meryl in 1970. The Sargents bought a house in the gentle wooded hills of Leverett, just north of the Amherst town line, where an ample piece of property full of oaks, hickories, birches, pines and hemlocks would serve as Sargent's outdoor laboratory. He was collecting *Catocala* every night of the summer season and had a trail of baited trees in his yard, which he checked at least four times a night. He installed four 150-watt outdoor spotlights and a 15-watt fluorescent blacklight, and his Robinson mercury vapour light trap was purring from dusk to dawn. This type of trap was designed by the eponymous Mr Robinson, one of whose claims to fame was to realize that moths are *not* drawn to flame. They are really trying to *avoid* lights, but a quirk in their visual and locomotory system causes them to see an illusory band of darkness around a light. The result is suttee by candle or porch light, celebrated in poetry as, among other things, the mystic soul's self-immolation in God.

Sargent's senses became exquisitely attuned to every detail of underwing life. Dark nights, he found, were better than bright nights; warm, humid nights better than cool, dry nights; still air better than wind. Light to moderate rain and fog were propitious for collecting. After ten years of intensive study, he was 'still convinced that the North American *Catocala* moths pose some of the most interesting and challenging problems to be found in evolutionary biology'. Because many of the underwings that came to his traps were black mutants, Sargent became intrigued with melanism.

E.B. Ford – austere, formal, radiating self-importance – came to Amherst College in 1966 at the invitation of Sargent's good friend, Lincoln Brower, and his then wife, Jane Van Zandt Brower. The Browers had spent two separate years, in 1957–8 and 1963–4, first as graduate then as postdoctoral students, in Henry's lab at Oxford. Jane Brower's classic work in butterfly mimicry became as much a textbook staple as Bernard's moths, and her intelligence, charm and patrician background made her one of the select few of 'female women' who rose above purdah in Henry's eyes. In Oxford Henry had been a kind and avuncular host, guiding the Browers around to tiny village churches and showing them the archeological treasures he and only he knew lay under a certain floormat, and whenever he came to the States he would spend several contented days with them. Their eldest son was his godson.

After delivering his lecture at Amherst College, Ford met with students in the very sort of American seminarish setting he abhorred. 'One of my students, a UMass grad student, was telling him about his research, which was really interesting from an evolutionary point of view,' Sargent recalls. 'Ford kept yawning. Then he said, "Do you mind if I sit down?" and he sat down.' Sargent also asked a question: How was it that there were so many melanic moths in rural areas where black moths were at a cryptic disadvantage? Ford answered dismissively: 'Heterosis. Next please.' This was Ford's stock answer to the question – heterozygous advantage explained how the melanic gene could hang on in rural areas. But Ford was wrong.

That night Sargent was invited to the dinner party at the Browers' palatial colonial house on the banks of the Connecticut River, at which Ford never ceased marvelling in his letters, sometimes listing dimensions in square feet. Sargent witnessed a supercilious Professor Ford who acted as if most others, including Sargent, were beneath his notice. 'These people have their ways of testing you, dropping a Latin or Greek phrase or mentioning something from opera. If you don't pass the test, you aren't the right kind.' After dinner, Sargent watched dumbstruck as Ford strolled out to the front porch, plucked a moth from the swarm around a light, and nonchalantly stuck it in his mouth. 'He said it was the only way to tell if they were palatable.'

It was at the 'LepSoc' (Lepidopterists' Society) annual meeting in Washington, DC in 1968 that Sargent met a florid and noisy Bernard Kettlewell. He struck Sargent as an affable, hard-drinking 'hail-fellow-well-met' type, as extroverted as Sargent was introverted, and in his talk he made several amateurish blunders in the way he discussed evolutionary theory. When Sargent stood up and asked a question, on a subject he knew very well, the black hindwings of underwing moths, he recalls that Kettlewell dismissed him with his 'disdainful' manner and crisp Oxford accent. 'He said I didn't understand what he was saying.'

Sargent understood quite well, and disagreed. His initial dissent centred on a pet theory of Bernard's, namely his conviction that peppered moths chose their own resting places for maximum crypsis. Before 'clamping down' on a surface, Kettlewell thought, the moths compared the background with the tufts of hair around their eyes. If the two were dissimilar, they experienced 'contrast/conflict' and adjusted their position. To test this hypothesis in 1954, he placed typicals and melanics overnight in a barrel lined with alternating strips of black and white paper; in the morning, he claimed, the moths had sorted themselves out by colour, pale moths resting on the white surfaces, black moths on the black.[180] In a subsequent experiment using tree trunks instead of paper strips, he again reported a difference in the background preference of the two morphs. However, to achieve his results he had to keep the males from flying away, which he managed by gumming together their fore- and hindwings, an unorthodox practice for a natural scientist.[181]

Sargent asked himself: Why should the moths prefer one background over another? If their resting sites were a matter of preference, both colour morphs would survive well in either polluted or unpolluted woods. Even in an unpolluted wood there are lots of dark places to sit. 'The background preference theory would predict that you'd always have black moths, because even without pollution there are always some black backgrounds – burnt areas, areas in shadow, diseased trees, some species of trees.'

In a series of experiments between 1965 and 1969, Sargent tried to replicate Kettlewell's background-preference work. He got contrary results, and concluded that the moths' resting places were genetically predetermined, not selected, as Kettlewell believed, by individual moths noting whether their 'circumocular tufts' matched the background. In one experiment, Sargent painted the dark moths white and the light moths dark. He even cut off their wings. In many cases *all* the moths preferred the white background. 'We tested many, many species, and among all the melanics we tested, the black moth always prefers white backgrounds. If I've done nothing else in my career I've demolished that hypothesis.'

Sargent's article 'Background selections of the melanic and pale forms of the cryptic moth *Phigalia titea* (Cramer)' appeared in *Nature*,[182] the world's leading science journal, in 1969 but was scarcely ever cited. Kettlewell and his compatriots acted as if the article, and its author, didn't exist, and his low visibility in the field nearly sank him when he came up for tenure at UMass the next year. 'I was young and had only a few papers, which were ignored. People in my department thought, "What on earth is interesting about light or dark moths choosing their background? I don't see any significance."' He got tenure, but only just.

In truth, the peppered moth was just one of Sargent's career problems, for when it came to academic politics he was his own worst enemy. So marginalized had he become by his retirement in 1999 that when, asked to write up his accomplishments, he reported having published nearly a hundred papers and having supervised thirty graduate students, including fifteen PhDs – all impressive numbers – his department did not believe him at first. When he offered to donate his moth collection of several thousand local specimens, UMass showed no interest, and the collection, worth thousands of dollars, went to a higher heaven: Yale University's Peabody Museum of Natural History.

One reason for his obscurity was that Sargent had refused to apply for grants. 'The pressure to get grants and renewals leads to lots of fraudulent data,' he explains. A grantless scientist, like a landless peasant in the Middle Ages, may become something of a nonperson, excluded from key committees at his or her own university and from scientific panels in the field beyond. There was a flip side, however. Having no grants, not being on anyone else's grant-review committee, Sargent could be said to be beholden to no one, and had no favours to receive or dispense. In the you-scratch-my-back-and-I'll-scratch-yours commerce of higher academia he was an incorruptible loner, a latter-day Ayn Rand hero who could say what he thought.

When a scientist publishes an article in a respected journal, he or she will order from the publisher hundreds of reprints to send to other scientists in the field who request them, and who may go on to cite this work in their articles. Scientists can network and keep abreast of one another's research in this way, and studies of the citation index have shown that future Nobel Prize winners can be predicted with uncanny accuracy by how frequently they are cited. Although Sargent was publishing respectably from 1969 on, he was stymied by the citation part of the cycle.

How much Sargent's invisibility was due to a habit of self-effacement, how much to his monklike distaste for groups and politics, how much to being ignored by Kettlewell, is hard to establish. His copy of Kettlewell's book bristles with exclamation marks, question marks, and aghast marginal comments in small, tidy cursive script ('horse shit' is inscribed next to one especially condescending statement). Kettlewell failed to mention the most important of Sargent's experiments, criticized the others, and claimed incorrectly that Sargent had never worked with a polymorphic species. He slighted Sargent further by leaving his major paper out of the bibliography and his name out of the index, which in academia signifies not just bad manners but something more akin to driving a disgraced member of an African village to his doom in the bush. Being frozen out by Ford and Kettlewell, and later by other scientists in the melanism field, would continue to undermine Sargent, even though his classic background experiments were cited by important evolutionary scientists, including Douglas Futuyma, author of the respected 1979 textbook *Evolutionary Biology*.[183]

As he examined Kettlewell's work more closely, Sargent began to suspect that not just the background-choice experiments but the whole classic story of industrial melanism might be off-base. The dark *swettaria* form of *Biston cognataria*, the North American peppered moth, was first recorded in Pennsylvania in 1906, later turning up in New Jersey in 1920, Chicago in 1935, and New York City in 1948. Its spread appeared to be as rapid as its British cousin's; in some industrial areas it had reached a saturation of 90 per cent by the early 1960s. Other species had melanic forms, too. So far, so good.

However, a number of lepidopterists working in the United States – David A. West, T.R. Manley, Denis Owen, and Sargent himself – had been consistently reporting high frequencies of melanic moths in very rural areas of Virginia, West Virginia, Pennsylvania and New England, a fact that seemed at odds with the 'changed background' hypothesis of industrial melanism.[184] Sargent was troubled by the fact that some of the melanics in his yard seemed too dark for crypsis, being darker than the darkest oak trunks. And why, he wondered, were they flourishing in a rural area, where the trees were covered with abundant lichen growth? In his 1976 book *Legion of Night*, he wrote:

One problem that poses a problem for a general explanation of recent melanism based on cryptic advantage is the extent of melanism in ostensibly rural areas (i.e., where the trees are not devoid of lichens and are not noticeably darkened by soot). My study area in central Massachusetts is an example, and here melanic frequencies are quite high in several species, including some *Catocala* . . . Many of these melanics are extremely dark, often nearly jet black, and would seem to be cryptic on only the blackest trees in heavily polluted areas. Furthermore, many of them, like their typical counterparts, prefer light backgrounds in experimental tests . . . making it even less likely that their occurrence is explicable in terms of a cryptic advantage.[185]

An accumulation of awkward findings in Britain also persuaded him that 'the incidence of melanism in some cases may be related to effects of industrialization other than observable environmental darkening.' Among these, he thought, were possible direct effects of pollution on the larvae 'through chemical contamination of the vegetation'. If these heretical words came to the notice of anyone in the industrial melanism field, they were ignored like the mutterings of a Hyde Park preacher.

I had been poring over Kettlewell's two original *Heredity* papers, which so many biologists and biology teachers had never read, and I was puzzled by a few things. What, I asked Ted Sargent one day in the local Starbucks, did he think was wrong with them? 'The original mark–release–recapture experiments were poorly done. They wouldn't hold up to statistical analysis. And Kettlewell scored the moths' crypsis according to aesthetic matching, which was subjective. He should have had blind observers, but he was the one who put the moths on the trees, scored their crypsis, and identified the ones that returned to his traps. You should be blind to the outcome of your experiments.'

I slapped the first paper on the table in front of us. Together we pondered the fine print of the various tables as if they were ancient runes. 'Look,' I said, pointing to Table 5. 'First there are just a few recaptures – one or two of each type – and then all of a sudden from July first on, he is getting a whole bunch. What can he be doing?'

'This is weird,' said Sargent, crinkling his brow. 'All of a sudden huge numbers of *carbonaria* are being recaptured. I've done a certain amount of release–recapture work, and the percentage of returns is usually much lower. I did what Kettlewell did with *Phigalia* and got very low returns and no difference in the ratios.'

There were oddities in the record of his Dorset experiment too. The first few days in June 1955, Kettlewell began recording the moths left on the trees at the end of each day, as he had done previously in Birmingham. Unfortunately he found more *carbonaria* survivors, where there should have been more typicals. He explained this by assuming the pale moths were *so* cryptic that he didn't see them. 'It became increasingly obvious that one was simply passing over the typical form on the lichened tree trunks, and they are practically impossible to see . . .' he wrote. 'For this reason, this type of recording was abandoned.'

He changed his methodology in mid-course again with his releases. On his first day in Dorset, he laid out 84 moths on 20 trees – about four moths per tree. He recaptured four *carbonaria* and one typical, quite the reverse of the results he wanted. The problem was that the moths were too highly concentrated, he decided; as soon as the birds discovered the smorgasbord they ate up everything in sight. All of this is confided in his paper, where he argued, furthermore: 'The birds in this wood were unlikely to have had any previous experience of the Black Peppered Moths, and it is conceivable, therefore, that at first some of them did not recognize them as an article of diet.' Maybe so, but it is hard to avoid the impression that

Bernard had a ready explanation for every unwelcome result and that when he got data he didn't like he often altered his experimental design.

His paper records that in Dorset he tried several variations of release technique until, on day four, he began releasing only two moths (one of each phenotype) per tree for the duration of the experiment. It can be argued that this is a reasonable density, given the sparse population of *Biston* in nature, but there had been no trace of this concern in Birmingham, where the high densities had yielded good results. It also means that his methodology in the two supposedly complementary experiments was different. Excluding three days' recapture results (when he was experimenting) improved his final ratio of survivors from 12.5 per cent of typicals and 6.3 per cent of melanics to 13.74 and 4.68 per cent respectively – a three to one advantage for pale moths.

'Kettlewell designed the experiment, he did it, he gathered the data. He knew the significance of the results and he got the results he wanted. We don't allow experiments like this any more.' The unspoken possibility of fraud hangs in the air. Sargent does not leap to that conclusion. 'It doesn't have to be fraud. There are subtle ways to seduce yourself. When I was a PhD student at Wisconsin I was raising zebra finches in nests made of different coloured strings. My theory was that the young birds would imprint on the string colour and use it later. One day [the eminent animal behaviourist] George Schaller said to me, "You know, when you change the cages you're really rough with some of them and gentle with others." He was right, and I hadn't noticed it. I was being rough with those cages where the birds weren't doing what I wanted them to do. I'd pull the door open, throw the food in there, jostle the cage a bit. When the birds were doing what I wanted, I'd gently ease the tray in.'

Unconscious bias, he says, could easily have tainted Kettlewell's background experiments, in which moths were placed in a barrel with black and white strips. 'The moth is a little too high, but you tell yourself, "Well, it's mainly on the black." If you have an idea in your mind it's very easy to eliminate what you don't want. Kettlewell always seemed to find what he expected would be true.' In the woods of Dorset and Birmingham, bias could have skewed the results in several ways. 'Who set the moths out?' Sargent asks. 'So many subtle things could enter. You might put the black ones closer to the collecting area, or hide the black ones a little better, unconsciously. Who recorded the data as the moths came back? In behavioural ecology work you have to hire blind observers to put them in the bag or boxes. It's easy to miss the paint mark and tell yourself, "That's not one of mine."'

He pauses, looks pained: 'His stuff is too neat! A two-to-one ratio and then two-to-one the other way, and nobody else gets anything like that? It's suspicious. It's exactly what you would have predicted from the theory. With any natural experiment in the field you don't expect to get results like these. In my yard a conspicuous moth that is there in the morning will still be there in the afternoon. Kettlewell's colleagues didn't want to shoot him down because they loved the idea. It was an example of Darwinism.'

In search of a second opinion, I prevailed on UMass biologist Donald Kroodsma to read over Kettlewell's two seminal papers and critique their experimental design. Not being a professional scientist I did not really know whether it was proper to keep changing horses in midstream as Kettlewell seemed to do.

'There is a fine line,' in the opinion of Kroodsma, who seemed sad about finding so much to criticize in another scientist's work, 'between "improving" experimental designs and adjusting them until one gets the desired results. It can be hard to know when one is "cheating".

'One should ask how it would have been possible for Kettlewell to falsify his hypothesis. If results didn't agree, he changed the methods until the data agreed with his ideas. With that kind of approach, it would have been impossible to falsify his hypothesis. And if we simply try to confirm hypotheses rather than test them, we're not really doing good science. He should have had blind observers and a more rigorous test, and the person who "knows" what the results ought to be should not be designing all the experiments.'

I had first become impressed with Sargent's powers of perception while researching the mysterious death of a student at U Mass, who had allegedly plunged through the roof of a university greenhouse and bled to death. Sargent has an office in the biology building adjacent to the greenhouse and sometimes tends the garden behind it. When he arrived the morning after the death, the campus police had already sanitized the death scene, removing all evidence of what had transpired, yet Sargent sensed that someone had died. The mockingbirds were singing a remarkably beautiful song, similar to the one they reportedly sang after the Battle of Gettysburg.

Sargent knew about the traditional relationship between mockingbirds and death from his extensive reading of nineteenth-century nature poets of New England. After noticing with dismay that the birds of his childhood were vanishing, he found that some of the best records of the songs and habits of certain nineteenth-century birds could be found in the poetry of Josiah Dean Canning (Gill, Massachusetts), William Cullen Bryant (Cummington, Massachusetts), and Amherst's Emily Dickinson, among others. 'The vesper sparrow is one. I knew it as a kid, a gentle bird that sings in the evening from the pastures. It was a favourite of the poets. Some said it was the bird Jesus would have liked. Now it's gone.'

As a bird expert as well as a moth man, Sargent was doubtful about the behaviour of the birds that ate Kettlewell's moths. 'Hedge sparrows don't eat moths off trees. Neither do robins. We've studied what happens to a bird's pursuit of a moth when it hits a tree. Most birds act as if the moth has just disappeared, as if it has gone to a different planet.

'Heslop Harrison said he had a wall behind his house with tons of moths and nothing would eat them; it didn't matter what they looked like,' he adds. 'At my house I see these moths that are just sitting here, so damn conspicuous. There are blue jays, nuthatches and woodpeckers around, and the moths are there for the whole day.'

One can, however, *train* a bird to pick a stationary moth off just about anything, including a tree trunk. Sargent has done it himself. Once, to illustrate how industrial melanism works, a scientific television show asked him to set up an experiment on the roof of UMass's Morrill biology building and place cryptic and non-cryptic moths on painted tree trunks. The raucous bluejays quickly wolfed down *every* moth in sight. The edited film shows the jays eating only the uncamouflaged moths, leaving the viewer with the impression that the camouflaged moths escaped unscathed.

'It was a demonstration, not an experiment,' says Sargent. In his eyes, Kettlewell's predation experiments were almost as phoney. 'He put out unnaturally high densities of the moths and they weren't in their proper resting places. What's the surprise that birds ate the ones they saw first? All you're really proving is that crypsis works, not that it's the reason for the frequency.'

It might be said that the birds in Tinbergen's famous film

were like conventioneers gorging on roast beef and shrimp at a buffet and leaving the aspic and stewed cabbages for later. Not since the initial critiques of P.B.M. Allan, Heslop Harrison and Bernard's peculiar old mentor, E.A. Cockayne, had anyone so attacked the central dogma of bird predation, and for a long time Sargent was as isolated as a desert prophet. 'When a theory becomes part of the common working knowledge of an entire community,' R.H. Brady, of the College of New Jersey, observed, 'it becomes the context with which that community understands the world. Doubt comes to be regarded as something less than legitimate, and critics find themselves only talking to each other.'[186] It is striking how early Sargent saw the flaws in evolutionary biology's most glamorous icon, a full decade at least before the scales fell from anyone else's eyes. However, he was not to be alone for ever.

Kauri Mikkola, a young researcher at the University of Helsinki who had been investigating melanism, took the train from London to spend a weekend in 1978 at Steeple Barton. Bernard Kettlewell was ailing, and his bad back kept him confined to bed for part of the day, but he watched eagerly from the window of his second-floor bedroom while Mikkola, outside on the lawn, went through the light trap, specimen by specimen, discussing the catch with him. Despite his infirmities, Bernard was a gracious and hearty host, and Mikkola was struck by the way he always addressed Hazel as 'my love' and 'my darling'. It was a tradition that a first-time visitor to Steeple Barton was served breakfast in bed on Sunday morning, but Mikkola slept through his alarm and was sound asleep when Nanny entered with his breakfast tray – a lapse for which he was teased for the rest of the day. As a special treat for Mikkola, Bernard had invited three great Oxford biologists, E.B. Ford, Niko Tinbergen and George Gradwell, to Sunday lunch. As everyone drank shot after shot of the rare 'Lakka' liqueur that Mikkola had brought, the distilled essence of golden cloudberries grown on the northern moors of Finland, the great men let their minds wander to rare caterpillars that feed on the cloudberry leaves. Awed by the august assembly, Mikkola politely refrained from mentioning his awkward finding, confirmed experimentally, that peppered moths did not rest on tree trunks. When he heard the news about Bernard's suicide the next spring, 'I was most happy that I did not argue with him about the resting background.'

In fact, Mikkola would have confirmed what Bernard had himself observed. 'Bernard knew this perfectly well,' Michael Majerus asserts. 'There is an obscure paper by Kettlewell on microhabitats in which he says he's often watched peppered moths take up their natural resting positions *on the underside of lateral branches.* He'd watched them doing this.' When his laboratory experiments showed that moths normally pass the day on the underside of branches, not on trunks, Mikkola pointed out that the conclusions drawn from the predation experiments could not be trusted. All the moths Kettlewell had recaptured on the first night following their release should be discarded, as they were sitting ducks where Kettlewell had positioned them, subject to unnatural predation. Kettlewell must have had his own nagging doubts, for he had bothered to keep a tally of moths caught 'after one day of self-determination', the ones, in other words, that had had a chance to migrate overnight and choose their own resting spots. Only these moths, Mikkola believed, could reveal the results of selection under natural conditions, and these did not add up to evidence of selective bird predation. If no selection had operated in Birmingham after the first night, the expected ratio of *carbonaria* to typicals would have been 37:6; the observed ratio was about the same – 37:5. New studies were urgently needed, Mikkola suggested, although it was several years before his paper was translated into English.[187]

This was one of the first public cracks in the standard model of industrial melanism, and it came alongside another, more theoretical attack. In 1982, the respected University of Wisconsin ethologist Jack Hailman, who specializes in evolutionary methodology, scrutinized 'this best-known case of evolutionary change' and asked an emperor's-new-clothes question: How do we *know* that the black moths replaced the light moths at Manchester due to selection? We don't know the total population, he observed, so perhaps the black moths simply 'became more numerous without effect on the light moth population'. Even if the population was shown to remain constant, he added, 'it is possible that one population can decline simultaneously with the rise of another population without there being a causal connection . . .' Kettlewell's results, he said, mix two separate findings – differences in survival of black and pale moths in release–recapture studies and differential predation by birds. It is nice to think that the first result is due to the second, but there is no empirical proof at all.[188]

In the same year, J.S. Jones of University College, London, announced in an influential article in *Nature* that the textbook case of industrial melanism was plagued by uncertainties: 'It appeared in industrial Britain not only in *Biston* and more than a hundred moths, but in species not subject to intensive predation such as beetles, cats, and birds.'[189] After pollution was curbed, melanics declined in many areas where they still appeared better camouflaged than typicals. A computer model by physicist G.S. Mani of Manchester showed that the geographical and temporal pattern of melanic frequencies in the British Isles could not be explained by crypsis alone.[190]

By the time a 1987 article entitled 'Exploding the myth of the melanic moth' appeared in the popular British science magazine *New Scientist*, the flaws in the case had multiplied. The author, Jeremy Cherfas, reported that 'after 20 years of moth-hunting' Rory Howlett and Michael Majerus of Cambridge had concluded that peppered moths generally rest in *unexposed* positions, in the shadow a few inches below a branch/trunk joint, on the underside of branches, or on twigs. Their study echoed Mikkola's findings from almost a decade earlier.

In fact, not only did the moths *not* rest on tree trunks – a finding corroborated a year later by Tony Liebert and Paul Brakefield in the Netherlands[191] – but a second crucial assumption was crumbling. As Cherfas reported, Howlett and Majerus took precise measurements of reflectance and reported that the light-coloured moths had partially translucent wings, so that 'in terms of the light reflected from them, they are more like black than white'. When the scientists performed Kettlewell-esque predation experiments with the moths in their proper resting spots, the dark moths were better hidden at the shaded branch joints than were the pale moths, and their camouflage was most effective in the *unpolluted* woods.

If typicals are cryptic on *dark* trunks, if black moths are well camouflaged in *unpolluted* woods, what is one to make of Kettlewell's classic experiments in Birmingham and Dorset?

Scientists in Britain had been trying to replicate Kettlewell's experimental results since the mid-Sixties, with mixed success. Predation and survey studies by Clarke and Sheppard (1966), Jim Bishop (1972) and David Lees and E.R. Creed (1975) presented a muddy and ambiguous picture, and later experiments, by Whittle *et al.* (1976), Steward (1977), Murray *et al.* (1980) and Howlett and Majerus (1987) did little to clarify it.[192] While some studies seemed to support the classical story, a number of other experiments reported no differential predation based on crypsis. Many studies revealed a gap between the predictions based on Kettlewell's model and the observations, about which Jim Bishop and Laurence Cook wrote in 1975: 'The discrepancy may indicate we are not correctly assessing the true nature of the resting sites of living moths when we are conducting experiments with dead ones. Alternatively, the assumption that natural selection is entirely due to selective predation by birds may be mistaken.'[193]

If selective predation by birds were the sole factor affecting frequency the most cryptic phenotype should have taken over in the areas where it was favoured, but this has never happened. Reviewing 115 years of breeding data, comprising 12,569 offspring from 83 broods, E.R. Creed *et al.* reported that unknown physiological factors in the pre-adult stages made the *carbonaria* homozygotes considerably hardier than typicals, heterozygotes, or the intermediate *insularia*. They analysed two cases where the *carbonaria* frequency increased rapidly in fairly unpolluted areas almost as soon as it appeared.[194] Clearly there was, in the words of J.S. Jones, 'more to melanism than meets the eye'.

'So much stuff has been published by British scientists since Kettlewell, and none of it is very conclusive,' is Sargent's conclusion. 'Everything came out so cleanly for Kettlewell – *too* cleanly – and no one since has gotten anything cleaner than mud.'

By the late 1980s, Laurence Cook at the University of Manchester was worried that the essential story was being buried under a morass of details. Cook first heard about peppered moths in 1955 when, as an undergraduate at University College London, he 'bumped into J.B.S. Haldane on the stairs saying that he had just come back from a *very boring* afternoon at the Royal Society'. Haldane had been listening to

Bernard unveiling his famous experiments. A few years later, when Cook was a graduate student in Ford's department, he went on a drive one day with Bernard, Hazel and young David to visit some tiger moth colonies founded by Bernard in the south of England. Along the way, Bernard stopped to pick out a nice, lichened 'unpolluted' log from Dean End Wood, Dorset, to exhibit at a Royal Society meeting. On the return trip, as Bernard's Plymouth swung around a sharp corner, 'I was nearly crushed to death by the log,' Cook remembers. Back at Oxford, Bernard cleaned up the log, and carefully placed the right moths on it. He called in the departmental photographer, John Haywood, and before long the 'tree trunk' that nearly dismembered Cook was in every textbook in the world. Students ever since have assumed that what they were seeing was a tree in an unpolluted forest naturally adorned with moths, not a log picked up in Dorset, transported in a Plymouth to an Oxford laboratory and decorated with moths. As the biology teacher Craig Holdrege observed in an article in *Whole Earth* magazine in the spring of 1999:

The impressive image of the camouflage of the peppered moth sticks in the mind, especially when the image is accompanied by a text . . . which gives no hint that we are looking at an artificially constructed situation . . . the explanation of industrial melanism appears in view of such images almost 'self evident.' The self-evident explanation dissolves when we learn that researchers don't find the moth during the day and that the pictures are composed by the researchers themselves.[195]

Cook had been doing seminal work on the peppered moth for more than a decade in the Manchester area. In 1989, with his colleague Mani, he arranged a meeting of the European peppered moth workers in Manchester, sponsored by the Linnean

Society. The scientists reviewed the evidence, including various geographic anomalies such as the high frequencies of melanics in rural East Anglia and the fact that after the Clean Air Acts were instituted, melanics decreased even in areas where typicals were still at a supposed visual disadvantage. They drew up a list of 'legitimate criticisms' of peppered moth research, including such experimental artifacts as excessive concentrations of moths, 'site selection assumptions naïveté', the use of 'incorrect resting positions (and use of dead specimens)'. The 'gaps in knowledge', they acknowledged, were many. 'It is clear,' R.J. Berry observed, 'that melanic peppered moth frequencies are determined by much more than differential visual predation by birds.'[196] Still, it was the consensus of the meeting that the basic model of industrial melanism was structurally sound.

By the early 1990s, if not before, it was known to a small circle of scientists that what every textbook in the Darwinian universe said about industrial melanism was untrue. There were some fundamental discrepancies, not least that birds may not be the major predators. The question is not whether a bird can be *trained* to eat a moth off a tree trunk – birds are known to be highly educable and the great tits in Bernard's aviary experiment in 1953 were 'quick to learn' from experience – but whether in nature birds are major predators of peppered moths.

Equally damaging to the 'authorized version' was the fact that moths do not normally rest on tree trunks. It is now universally acknowledged that Cyril Clarke, who laconically observed that in twenty-five years he had seen exactly two *Biston* resting on tree trunks, was right after all: the normal daytime resting place of peppered moths is *not* on tree trunks but in shaded areas under branches, where colour differences would be muted. According to Majerus: 'If the relative fitness of the morphs of the peppered moth does depend on their crypsis, the resting position is crucially important to the estimation of fitness differences between the morphs.'[197]

Additionally, the experimental densities were too high. In nature peppered moths are known to be very scantily distributed, but Bernard set out at least four moths per tree, and then replaced them as soon as all of one type were eaten. When he and Tinbergen were making their historic film, they laid the spread on even thicker. Everyone now concedes that these densities were unnatural. Kettlewell was, in effect, creating a feeding tray, and the 'intensity of predation' recorded in his experiments simply reflected a learned response by the local birds.

Furthermore, the method of release was faulty. Peppered moths fly at night and settle into their daytime resting places at dawn. Bernard released his moths in daylight because if he had released them at night they would have made a beeline for the light traps. ('I admit that, under their own choice, many would have taken up position higher in the trees,' he admits in his 1955 *Heredity* paper, '[and] in so doing would have avoided concentrations such as I produced.') One morning he tried releasing them just around dawn, but this proved too laborious; they were so cold he had to warm them up over his car engine to get them going. Did this method of releasing moths in daytime, when they are dopey and sluggish, and then placing them on trunks, make them abnormally vulnerable? Mikkola, for one, observed that this method doomed the moths to 'atypical' positions. 'In my view,' wrote Bruce Grant, of the College of William and Mary in Virginia, 'the greatest weakness in Kettlewell's mark–release–recapture experiments is that he released his moths during daylight hours . . . It would have been much better had he released marked moths after dark on nights when traps were intentionally shut off.'[198] Grant once dramatically displayed the moth's diurnal somnolence at a scientific meeting by placing live peppered moths on a beer can, where they rested contentedly throughout the proceedings.

In view of all this – stuporous moths placed in unnatural resting sites in unusual densities – the bird predation purportedly demonstrated by Kettlewell could easily have been an artifact of the experiment. One critic referred to the situation as 'unnatural selection'.

The flaws didn't end with the method of release: there was also the question of what Bernard put out. He mixed lab-bred and wild-caught specimens and released them in the woods. The numbers were unequal. For obvious reasons, the phenotype that was in low frequency at the site had to be supplemented with laboratory moths, so that in Birmingham many more of the typicals were lab-bred, and in Dorset virtually all of the *carbonaria* were. 'This was a mistake,' asserts Bruce Grant. 'Kettlewell should have tested for differences.' Noting that the lab-bred moths were possibly 'more vulnerable', Jerry Coyne of the University of Chicago points out that in each place, the rarer, and supposedly less cryptic, phenotype would have been lab-bred. If it did not survive as well, it could have been for reasons other than camouflage.[199]

Then there is the issue of bird vision. Bernard's experiments implicitly assumed that what is cryptic to the human eye would also be cryptic to a bird. Yet since the 1980s it has been known that bird vision and human vision are quite different. Birds have greater visual acuity and colour discrimination than we, and, with their additional vision 'cone', can see well into the ultraviolet spectrum, discriminating different wavelengths in the UV range. One scientist, Jim Stalker, reported that while to the human eye black moths were more conspicuous on foliose lichens, the reverse was true in the U V spectrum perceived by birds. Majerus hazards the suggestion that peppered moths are adapted to *crustose* lichens instead, but he concedes that 'none of the assessment of the relative crypsis of moths as determined by humans should be applied to birds'.[200] Moths look different if you are a bird.

Another body-blow to Kettlewell's hypothesis is the absence of any proof showing that bird predation depends on crypsis. Kettlewell's experiments supposedly showed that the less chance a moth has of finding a resting site where it is camouflaged, the more likely it is to be eaten by a bird. This assumption could be said to be the bedrock of the accepted peppered moth story. 'Yet, surprisingly,' notes Majerus, 'experiments to show formally that the degree of crypsis of the different peppered moth forms does affect the level of predation inflicted upon them by birds have never been carried out.'[201] Jack Hailman had made a similar point. There was no evidence.

And the contradictory evidence continues to mount. Lichens figure so prominently in the textbook model of industrial melanism — all those wonderful pictures — that it can fairly be asked if without lichens we have a case. David Lees' and Robert Creed's 1975 study demonstrated that melanic peppered moths thrived in some rural places like East Anglia, where the trees are lichened and the atmosphere unpolluted. This led them to infer the presence of a 'counteracting selective agent which may occur in any stage of development and may be only indirectly related to the colour of the adult moth'.[202] In North America, the rise of black moths did not depend on an absence of lichens, and in Britain the pale moths returned before the lichens did.

Even the function of melanism remains unclear, and one cannot assume that industrial melanism arose because of cryptic value. Melanism serves as warning coloration in some insects, and as a defence against ultraviolet radiation in other species, including man. Melanic pigeons fared well in towns and cities because associated hormonal changes allow them to breed in wintertime. The blackened form of the two-spot ladybird beetle (*Adalia bipunctata*) absorbs solar energy more efficiently. Black forms of the pale brindled beauty *Phigalia pilosaria* in Britain and the closely related *Phigalia titea*, or half-wing geometer, in the US are considered industrial melanics yet show no clear relationship to predation, nor have predation studies turned up differential survival among the pale and melanic morphs of the scalloped hazel, *Gonodontis bidentata*, which Bernard once contemplated using in his experiments. 'The variety of types of associations of melanism with air pollution, and the large number of species involved in various parts of the world, make impossible generalisations about the facts, selective or otherwise, which influence industrial melanic polymorphisms,' David Lees wrote in 1981.[203]

And what about the caterpillars? Kettlewell knew that mortality among peppered moths (and all moths) occurs overwhelmingly before the adult stage. Eggs, larvae and pupae perish in nature at a rate of at least 90 per cent. Moths on the wing are only a small band of lucky survivors. 'If you did a life table of peppered moths,' Majerus reckons, 'I would be very surprised if you got ten per cent of eggs that were laid surviving to pupae.' Another lepidopterist estimated the percentage at 1 to 3 per cent. This doesn't affect the theory of industrial melanism as long as egg, larval and pupal mortality isn't *differential* mortality – which Majerus assumes is the case – but the truth is that we know very little about the natural history of the peppered moth. Focusing exclusively on avian predation of adult resting moths seems a bit like determining which law students will pass the bar by examining their clothing during the final week of classes and ignoring study habits, exams, and brief-writing ability. In

1980 Robert Creed and David Lees determined from breeding records that physiological events during the pre-adult phases appeared to be crucial, giving melanics the edge.[204] Lees told me that larval characteristics, for example, may explain why there are more typicals or more *carbonaria* in a given place in a given year. The unsettling implication is that we may be seeing the outcome of selection on the larvae – and adult differences may be random.

Bats further complicate the picture: Kettlewell himself admitted that they probably accounted for 90 per cent of the predation of adult moths. That didn't matter, he always insisted, because bat predation wasn't *differential* predation; evolution was driven by the small percentage of moths that are eaten selectively by birds hunting visually.[205] Crypsis could have little or no bearing on bat predation since sonar, rather than sight, is used to locate their prey.

It's important to be precise about what is mortality and what is differential mortality. Majerus defines it by example: 'Say three hundred eggs are originally laid. Once you get to the adult stage, maybe you have ten left. Of these more than half are killed by things not hunting by sight, so say you have four moths left – two typical and two *carbonaria*. You must be prepared to say that none of the mortality prior to this is due to selection on colour pattern, no pleiotropic effects of alleles, no differences in palatability, no greater energetic costs in producing black pigment and so on. If so, then despite 296 moths being killed up to that point, if those two typicals are eaten by birds, you've increased *carbonaria* by a hundred per cent at one go.'

The objections to this assumption are similar to the objections to ignoring pre-adult mortality. Can we really be sure that bat predation is *not* selective, that there is not some yet unidentified difference between melanics and typicals that makes one morph more vulnerable to bats? Certain night-flying moths can dodge or jam bat sonar, according to several studies, and it is not known whether this ability is equally distributed.

With all these defects, one might suppose that 'Darwin's missing evidence' had lost a little of its lustre. On the contrary: the leading lights of the industrial melanism field still cling to it, even while admitting that there are all sorts of confounding factors.

'I know from my own work that very, very rarely can you put a single effect down to a single cause,' Berry tells me. 'What you measure at the end is the frequency of a gene, but to say that it is caused by, say, the distribution of lichens, is a gross oversimplification. You would have physiological factors, growth factors, how hot and cold the air is, how hungry the birds are. You may have toxic factors, what they're eating and whether this affects the metabolism. But to say that these factors are important shouldn't distract from the fact there is a simple story in the background. Kettlewell's work hasn't been disproved. The black form is clearly an adaptation.'

Everyone – in Britain, anyway – believes this as firmly as they believe the sun will rise tomorrow, just as they consider bird predation an incontestable fact. 'It's very clear that the birds did eat the moths, and the more conspicuous ones first,' says Berry. 'So there's no question about that.' Based on Tinbergen's film and other reports, Majerus too feels certain that birds eat 'significant numbers of moths'. He admits, however, that there is no proof, that 'observations of peppered moths being taken from natural resting positions are still lacking and are urgently needed'. Without bird predation the textbook case would lose much of its punch.

*   *   *

On 12 November 1996 the peppered moth had its Warholian fifteen minutes on the front page of *The New York Times* science section, in a report that, some would say, undermined one of the main pillars of the traditional model.

Professor Bruce Grant of the College of William and Mary in Virginia had been assigning his graduate class Kettlewell's papers to read each year, and every year when they got to the background-matching papers Grant wondered anew about the phenomenon. 'It was a great idea. It meant that if a new melanic form arose, it would automatically do the right thing. Sadly, it wasn't true.' Like Kettlewell, Grant thought that there was a gene that coded for site-selecting behaviour in the peppered moth and that it might be co-adapted, or harmoniously combined with, the gene for colour. Like Kettlewell, he dismissed Sargent's papers on the subject, some of which he hadn't read. He decided to try to repeat Bernard's experiment with *Biston cognataria*, the American version, but he couldn't catch enough. Colleagues told him that the man to see was Cyril Clarke, who 'trapped more moths in one summer than anyone else altogether', and soon he was ensconced at Clarke's estate on the Wirral Peninsula near North Wales, building moth pens with black and white stripes. 'I thought, if I open this in the morning and the black moths are sitting on the black stripes, and the white moths are on the white stripes, that's going to be mind-blowing.' But Grant went through a series of test pens, and also fashioned miniature Elizabethan moth collars to fool black moths into thinking they were white, and vice-versa, without finding a shred of evidence for the 'behavioural polymorphism' so sacred to Bernard Kettlewell. 'I did find that some moths had individual preferences, but they weren't related to their colour.'

Now he was beguiled by the peppered moth, and he teamed up with the genial Sir Cyril and his wife Frieda, who had taught herself genetics and performed most if not all the laborious breeding work for her more butter-fingered husband, who tended to leave the rugs littered with dropped insect pins. Sir Cyril and Philip Sheppard had both referred to Lady Clarke as 'our technician', and she was called in whenever her husband needed to focus a microscope. 'I don't think Cyril would have had the patience for the breeding work,' one acquaintance mused. 'You have all these caterpillars that look exactly alike, and you have to keep them separate and know what strain they are.' From 1959 until the late 1990s, the Clarkes never missed a summer of daily trapping and ended up with data spanning more than three decades, more than 18,000 specimens in all, that documented in full the historic demise of the melanic peppered moth.

Since clean air acts began to be implemented in 1956, pollution levels, as measured by sulphur dioxide and smoke, had been dropping steadily in Britain. In the 1980s, three different surveys documented an extraordinarily rapid decline in the frequency of black peppered moths, from 90 per cent in industrial areas in the 1960s to as low as 40 per cent by the mid- to late 1980s. In 1994 the frequency of *carbonaria* in Cyril Clarke's garden had reached a low of 20 per cent. The dark moths were disappearing as swiftly as they had appeared, a fact that seemed to lend credence to the textbook story of industrial melanism.[206]

In the United States, where clean air laws were enacted in 1963, a parallel process was under way. Black forms of *Biston cognataria* were initially rare in southeastern Michigan. Denis F. Owen, an eccentric, hard-driven, chain-smoking British scientist from Oxford Brookes University, knew that from looking through local museum collections. When Owen trapped peppered moths in the George Reserve near Detroit from 1959 to 1962, he got 90 per cent melanics; other collections from that era were similar. 'There was drawer after drawer of moths, hundreds of them, and they were all black,' Bruce Grant would recall. In 1994–5, Grant hooked up with Owen, to whom he was introduced by Cyril Clarke, and revisited the Reserve. Now the melanics constituted just 20 per cent of the population.

This 'parallel evolution' seemed a Darwinian miracle, noteworthy enough to inspire *The Times* to run a rather breathless article, 'Parallel Plots in Classic of Evolution', about the paper that Grant, Clarke and Owen published in *The Journal of Heredity*.[207] The evolutionary biologists quoted in *The Times* sounded as thrilled as astronauts walking on the moon. 'When you have parallel changes like this,' one said, 'it's like having different replicates in an experiment. The more you have, the more confident you are that you're getting a consistent result.'[208]

The dramatic rise and fall of melanism in the Michigan forest had all taken place in the midst of luxuriant lichen growth; the lichens had never gone away. In Clarke's backyard – as well as another site he studied – the lichens were gone, but the typical peppered moths began returning *before* the lichens came back in earnest, a fact he noted emphatically in papers in 1985 and 1994. The traditional story of industrial melanism held that industrial fallout destroyed the moths' lichened backgrounds, but Grant, Owen and Clarke wrote: 'the changes in allele frequencies in the moth populations we sampled have occurred in the absence of perceptible changes in the local lichen floras. We suggest that the role of lichens has been inappropriately emphasized in chronicles about the evolution of melanism in peppered moths.' The critical factor must be atmospheric sulphur dioxide, the authors concluded, for improvements in Michigan's air quality 'parallel the changes recorded in Britain'. Previously everyone had assumed that sulphur dioxide had exerted its effects by killing lichens on the trees, but Grant *et al.* suggested that it might work in some other, unspecified way – thereby possibly making coloration beside the point.

It was a sign of the times that the author of the *Times* story, Carol Kaesuk Yoon, referred to the known flaws in the textbook model and stated that, instead of acting through birds, 'natural selection might be acting on the caterpillars destined to become light or dark moths.' She added: 'While it is unclear exactly how natural selection is acting, it is very clear that natural selection is hard at work.'

In Britain the attack on the lichen factor met with a fairly cool reception. Michael Majerus dismissed Clarke's lichen assessments as 'anecdotal',[209] while Laurence Cook insisted that the lichen picture was more complex. He and his colleagues had surveyed a 125-kilometre transect from polluted Manchester southward into unpolluted North Wales in 1973 and again in 1986. They reported that despite an overall decline in lichen area and diversity, the sites that had formerly been the most polluted had actually gained in lichen cover by 1986 – and that this corresponded to increases in the frequencies of typical peppered moths.[210] Still, the observation that the typical moths came back before the lichens seems undeniable.

The correlation between sulphur dioxide and melanism, furthermore, sounds convincing, but does it really tell us anything? During this period of history sulphur dioxide increased and then decreased almost everywhere. As Sargent points out, 'You could correlate it with anything – VCRs or minivans or computers. Where is the other half of the experiment?' If there is a link to sulphur dioxide, then areas in which sulphur dioxide increased or stayed constant should continue to be dominated by melanics. However, Grant admitted that he has sampled *Biston cognataria* in the highly polluted environs of Weirton, West Virginia, a steel town west of Pittsburgh, and found the black moth population down to less than 6 per cent, even lower than in many unpolluted forests.[211]

Grant is not bothered by what he sees as a small discrepancy. To him the corpus of research on the peppered moth is 'the largest single record documenting an evolutionary change observed in any species, and natural selection is the only force known to science that can explain it. Evolution is defined as a change of allele frequency in time. This we know. We're getting one generation per year; when we follow it for fifty generations we have a substitution. That cannot be explained by genetic drift or migration or any other evolutionary force except natural selection. Only natural selection can explain the steady trajectory and the speed.'

He is not alone in insisting that the peppered moth continues to embody the noble truths of evolution. No matter how flawed, the basic message continues to be broadcast that one factor, avian predation on resting moths, is effecting the changes in gene frequencies. Other factors may be invoked ad hoc to explain discrepancies, but they are treated as mere details, akin to errors in spelling or punctuation. The accumulation of fifty years' research has built a monument that researchers in the field are understandably loath to dismantle. The worst-case scenarios, such as the possibility that the rise of melanic peppered moths may not have demonstrated natural selection, are unthinkable. Only slightly less disturbing is the possibility that natural selection is operating at the little-understood pre-adult stage, when the adult wing colours are still concealed from selection and the key element of crypsis in relation to environmental change would not apply. If the major predators should turn out to be bats or beetles, instead of birds hunting by sight, or alternatively if the birds are picking the moths out of the air, the standard model is in trouble again. But almost no one really wants to re-examine the theory itself. Those few who do are demonized.

12

A damn good story

I t is hard to imagine a place less like Oxford's mist-wreathed spires than the University of Massachusetts at Amherst. It is part of the school's strange karma that it owes much to the legacy of Lotta Crabtree, a popular and wealthy nineteenth-century actress and comedienne with an excessive fondness for animals. In 1924 the 'queen of the American stage' willed $1.25 million for interest-free loans to U Mass graduates, but the bulk of her $4 million estate was left to the Dumb Animal Fund to provide drinking fountains for dogs, cats and horses all over the country. Not satisfied with its portion, the University of Massachusetts successfully contested Crabtree's will, on the grounds that the pet-drinking-fountain scheme was bonkers, and pocketed the whole sum.

Today, U Mass/Amherst, the largest of the five University of

Massachusetts campuses, is an enormous institution of 24,000 students, the sort of undergraduate horde that E.B. Ford would have deplored. It suffers, many say, from an inferiority complex fostered by sharing the same state with Harvard and MIT, with which it seems to compete not so much in scholarship as with concrete and steel. The university is a shocking sight to the motorist driving north on Route 116 through the fertile, sweet-hay-smelling farmland of Hadley. Its charmless utilitarian towers rise out of the floor of the green, once drowsy valley like a cut-rate Brasilia, a skyscraper movie-set fused with some spare scenery from *North by Northwest*. In a town whose bylaws prohibit any structure higher than four stories, the W.E.B. DuBois library tops out at 28 stories. (To browse among the biology and entomology journals, it was necessary to take an express elevator up to the 23rd floor; the nearest copy machine was still higher, on floor 26.) The five high-rises of the Southwest dormitory complex, housing eight to ten thousand students, tower to 22 stories, and more than one 'elevator-surfing' undergraduate has fallen to his death. There isn't a cat drinking fountain in sight.

Much of the building boom occurred in the 1970s and 1980s, as the former agricultural school transformed itself into a large state university with pretensions. Those buildings and aspirations proved expensive for state taxpayers, and now two-thirds of the university's income comes from 'other sources', that is, grants. The mindset of the nouveau regime was communicated in a terse mid-1970s memo from the dean of the School of Arts and Sciences ranking the faculty by grant money earned and office space occupied. 'We were rated in dollars per square foot. Anyone low on grant money kept having office space taken away,' recalls Sargent, who was, of course, grantless. After a fairly idyllic decade and a half spent quietly devising testing apparatus for moths and writing articles that were published in

Science, Nature, Evolution, and other respectable journals, he found himself so devalued at his own university that students in the prestigious new Neurosciences and Behavior department could not get credit for taking his courses in animal behaviour. He was president of the Lepidopterists' Society and editor of its journal, but in 1975 UMass could not cough up the funds to send him to Alaska to deliver the presidential address.

'In the last five years here I couldn't get students because the only way to get students was to go through committees, and they give students to "active labs" – with grant money. The man who hired me thirty years earlier had said, "This is one of our great researchers, and he's doing his research in his own backyard."'

In his own backyard, in fact, on 23 May 1986 Sargent caught a black moth in his light trap. Recognizing her as a female melanic *Panthea pallescens*, he put her in a plastic jar with a small sprig of white pine, the larval foodplant, cut from a tree in his yard. She obliged by laying eggs at once. Three days after they hatched, Sargent separated the tiny, naked-looking caterpillars into two separate plastic ice-cream pints, and raised one group on new-growth pine needles (the bright-green needles of the current year) and the other on old-growth pine (of previous years).[212]

He soon noticed that the caterpillars eating the new-growth pine were growing more slowly than the ones in the other container, and they pupated and eclosed later. In both groups all the male moths were melanics, and among females overall there was a fifty–fifty ratio of typicals and melanics. This meant that, assuming the melanism was controlled by a sex-linked dominant allele, the melanic female had mated with a heterozygous melanic male. However, there was a statistically significant difference between the two groups. More than two-thirds of new-growth-fed female moths were melanic, while two-thirds of the group fed on old-growth needles were typical. A second experiment on 8 July yielded similar results. Sargent believed he had hit pay dirt.

'Something in the new-growth needles was favouring the expression of adult melanism,' Sargent explains. 'Some kind of induction was going on.' In this context, the word *induction* is highly charged, a Molotov cocktail of a concept. Back in the 1920s Heslop Harrison claimed to have induced melanism in some moth species by feeding them on leaves coated with industrial pollutants and – here is the dicey part – purportedly demonstrated that the environmentally induced trait was *inherited*. After Heslop Harrison was discredited, this idea fell out of favour like phlogiston or the flat earth, due in no small part to E.B. Ford's crusade against Heslop Harrison. Citing Heslop Harrison in papers, it could be argued, was not a smart political move, but Sargent did it anyway.

'Mutations occur by chance,' he muses. 'According to the classical story, there was a chance coincidence of darkened environments and a gene that would make you darker. That seems unlikely in so many species – at least seventy species in North America.' The rise and fall of melanism happened *too* fast, suspiciously fast, he thinks. Peppered moth populations around Manchester, England, went from zero melanics to more than 95 per cent melanics in a mere forty-seven years, between 1848 and 1895. This shift, if due to a change in the frequency of a dominant allele, requires a 50 per cent selective advantage of the melanic over the typical morph – a figure that E.B. Ford seized on greedily in his determination to show that selective pressures could be very powerful in nature.

'No one has really questioned the speed of the process,' Sargent insists intently. 'It's like no other process in nature we've ever encountered. In this area, among *Panthea*, melanics went from zero to seventy-five per cent in seventy-five years.

If there had been any melanics around in 1903 they would have been in Holland's *Moth Book*, and they are not. There were none in my traps when I was a kid sampling. Then, suddenly, there were many melanics, in many different species. You might say, "My God, what an incredible selective advantage!" But was it? It strains the idea of chance.

'My own belief is that there is something funny about melanism. I think of the subtitle of Kettlewell's book – "a recurring necessity". That it happened so fast makes me think the genes were always present and were turned on by something in the environment, and now they're being turned off.'

Another anomaly he sees is this: in some species he has tracked from year to year, such as *Phigalia titea*, the percentage of melanics has stayed at 20 per cent for twenty years. 'It's still twenty per cent. It's twenty per cent in my backyard, it's twenty per cent in New Haven, it's twenty per cent on Cape Cod. By chance it *shouldn't* have occurred in some places if it's due to natural selection working on random mutation.'

Although Sargent would undoubtedly be described by nine out of ten eyewitnesses as 'quiet and unassuming' – a mild, grandfatherly figure frequently overlooked by waiters in restaurants – he is a dangerous iconoclast in the eyes of the industrial melanism establishment. He finally published, with two co-authors, a devastating analysis of the classic industrial melanism story in 1998,[213] concluding that 'there is little persuasive evidence, in the form of rigorous and replicated observations and experiments, to support [the classical] explanation at the present time.' Although it enraged the community of his peers – Bruce Grant called it a 'dreadful review' and a 'hatchet job' – Sargent's article was *not* the decisive confrontation of the peppered moth wars. That erupted in the 5 November 1998 issue of *Nature*, in a review written by Jerry A. Coyne, professor of ecology and evolution at the University of Chicago, of a new book by Michael E.N. Majerus. The book, called *Melanism: Evolution in Action*, was a watershed event. Methodically and incisively analysing every flaw in Kettlewell's experiments and in the industrial melanism paradigm, Majerus's book left no doubt that the classic story was wrong in almost every detail. Peppered moths, if left to their own devices, surely do not rest on tree trunks; bird vision is nothing like human vision; Kettlewell was wrong about how peppered moths choose their resting sites; the high densities of moths he used may have skewed the results; the method of release was faulty, and on and on. The various predation and survey studies conducted after Kettlewell have not replicated his results particularly well, and other 'factors' kept having to be invoked to squeeze the data into the standard industrial melanism model. 'The findings of [scientists since Kettlewell],' Majerus concluded, 'show that the precised description of the basic peppered moth story is wrong, inaccurate, or incomplete, with respect to most of the component parts.'

The reader who makes his way through Majerus's mountains of evidence is rather stunned to arrive at his verdict: that the basic story, while 'undoubtedly more complex and fascinating than most biology textbooks have space to relate', is perfectly fine. 'My view of the rise and fall of the melanic peppered moth is that differential bird predation in more or less polluted regions, together with migration, are primarily responsible, almost to the exclusion of other factors.'

Jerry Coyne, however, was 'horrified'.[214] The sheer magnitude of the problems itemized in the book filled him with dismay and something like shame. After all, he too had been teaching the 'standard Biston story' for years. When he dug out Kettlewell's original papers he found that things were even worse than he thought. How was it that the experiment that Coyne called the 'prize horse in our stable of examples' had been accepted unquestioningly all this time? Was it possible that the facts had been submerged because 'such powerful stories discourage close scrutiny'? Concluding that 'we must discard *Biston* as a well-understood example of natural selection in action, although it is clearly a case of evolution,' he mused:

B. betularia shows the footprint of natural selection but we have not yet seen the feet. Majerus finds some solace in his analysis, claiming that the true story is likely to be more complex and interesting, but one senses that he is making a virtue of necessity. My own reaction resembles the dismay attending my discovery, at the age of six, that it was my father and not Santa who brought the presents on Christmas Eve.[215]

Not so much because of Majerus's book as because of one review of it – especially the felicitous phrase about Santa Claus – the paragon of natural selection was outed. It was not long before sceptical articles popped up in newspapers and creationist tracts, some written in the gloating tone used by tabloids when they uncover the 'real truth' about Madonna or O.J. or Princess Di. After summarizing the latest findings about peppered moths' natural resting places in a 1999 article in *The Scientist*, biologist Jonathan Wells, a fellow of Seattle's Discovery Institute, a 'scientific creationism' think tank, quipped: 'It seems that the classical example of natural selection is actually an example of unnatural selection.'[216] A widespread, gleeful creationist attack on the peppered moth went into high gear. At the same time, a 1999 article by Robert Matthews in the London *Daily Telegraph* began:

Evolution experts are quietly admitting that one of their most cherished examples of Charles Darwin's theory, the rise and fall of the peppered moth, is based on a series of scientific blunders.

Experiments using the moth in the 1950s and long believed to prove the truth of natural selection are now thought to be worthless, having been designed to come up with the 'right' answer.

Scientists now admit that they do not know the real explanation for the fate of Biston betularia, whose story is recounted in almost every textbook on evolution.[217]

And so on. Kettlewell's experiments were called 'essentially useless', and Jerry Coyne was quoted as saying: 'There is a lot of wishful thinking and design flaws in them, and they wouldn't get published today.'

As portions of that article spread like an equatorial virus through newspapers and anti-evolution websites around the world, many biologists and biology teachers felt unmoored. Instead of a peerless example of the workings of microevolution, was 'Darwin's missing evidence' just an empty demonstration, a red-faced wino in a Santa suit? 'We don't always read the original papers,' admits Douglas Futuyma,[218] of the State University of New York at Stony Brook, who a few years earlier had been quoted exulting over 'parallel evolution' in *The New York Times*. 'It's clear that there is much more going on here than bird attacks and camouflage.' Majerus began receiving panicked e-mails from biology teachers in Canada, Australia, Brazil and the US asking if they could still use the example. 'Without exception, I say, "Yes, you can,"' he tells me. 'And I say Jerry Coyne tried to put words in my mouth. He wanted to say it is not a safe example of Darwinian evolution. I'm quite happy if he feels like that but I would object to his suggesting that I feel like that.'

Battle lines were being drawn. Critics of the peppered moth experiments who nonetheless saluted the basic model

– Majerus, Mikkola and Bruce Grant, for example – were still accepted in the industrial melanism club, but Sargent and Coyne were clearly beyond the pale. They had violated a major taboo by suggesting that bird predation, the *sine qua non* of the story, was unproven, and that natural selection was not necessarily nailed down in this case either. Both men have been accused, in direct and indirect ways, of giving aid and comfort to the enemy, the creationists. When they are cited in papers it is often, though not universally, dismissively. In Bruce Grant's papers, especially, Sargent lurks as a deluded malcontent, 'subverting' the classical story and reviving discredited induction arguments 'as if they were worthy of serious consideration'. Grant also repeats Kettlewell's contention that Sargent never worked with polymorphic species in his background experiments, even though Sargent's 1969 paper in *Nature* was a test of the background selections of 'the pale and melanic forms of the cryptic moth, *Phigalia*', to quote the title. 'These charges make me look like a nut,' is Sargent's complaint.

'They think I'm a Lamarckian,' he adds, 'though they might not use the word.'

At its *reductio ad absurdum*, Lamarck's theory of use–disuse predicts that if you practise the guitar a great deal, your children will be born with calluses on their fingertips and better than average strumming ability. And the 99-pound weakling who bulks up on the Charles Atlas programme will have ultra-steroidal pumped up sons. It is an attractive idea to some because it provides a mechanism for purposeful evolution, in contrast to the dismaying purposelessness of natural selection. For some reason Lamarckian ideas have often had a tumultuous, if not fatal, effect on the scientists who indulge them. A British scientist named Paul Kammerer who had tried to prove the inheritance of acquired characteristics in certain amphibians committed suicide after being accused of fraud in *Nature* in 1920. The writer Arthur Koestler, among others, believed that Kammerer had been falsely accused and wrote a book about the case called *The Midwife Toad*. But Sargent is talking about something rather different.

The induction of various colour traits by environmental pressures – such as temperature – is a well-known phenomenon in insects and other organisms. Siamese cats and some rabbits possess 'heat-shock' genes that turn the tips of their ears and their extremities black in high temperatures. Seasonal forms in many butterflies and moths are controlled by such genes as well. The gene is always there but it is expressed only under certain conditions. Some of Bernard Kettlewell's initial correspondence with his future boss, E.B. Ford, concerns experiments in which Ford heated the pupae of *Heliothis peltigera* and produced uniformly melanic adults.[219] Kettlewell himself induced melanism in some moth species by rearing the larvae at higher temperatures. According to Lincoln Brower, monarch butterflies can turn melanic if as caterpillars they feed on toxic milkweed. Induction of melanism by crowding is common in moths and grasshoppers: crowd the larvae and you get black forms.

However, countless breeding experiments with peppered moths have produced the expected Mendelian ratios in the offspring. Thus, the melanic trait is demonstrably genetic, and if it is genetic, how can it be induced? Laurence Cook flatly rejects the induction theory. 'It's absurd'.

Sargent argues that an induced trait may *seem* to be inherited in a Mendelian fashion if what is actually inherited is a sensitivity to the inducing agent. He conceives of an allele that is turned on or off – expressed or inhibited – in the presence of a particular environmental pressure. 'This idea is not crazy,' avers John A. Endler, professor of ecology, evolution and marine biology at the University of California at Santa Barbara.[220]

If melanism is not a response to a general darkening of trees caused by airborne pollutants, as assumed by the classical model, it must be a response to some other rather recent ecological change, Sargent believes. In New England, he proposes, it may be a byproduct of the relatively recent conversion of large tracts of agricultural land to woodlands. Most of the virgin forests that the colonial settlers first encountered were cleared for farmland by 1750. As agricultural lands were abandoned in more recent times, these areas have been recolonized by new, different tree species. In contrast to the older trees such as beeches, maples, oaks and hickories, which have light bark, the recolonized transitional woodlands tend to be filled with trees with dark bark (cherries) or variegated black and white bark (white birches and poplars).[221] Sargent's experiments have suggested that melanic moths could be mistaken for the dark patches on birch trunks. Philip Sheppard found years ago that melanic moths seemed to be cryptic on birch, and Bruce Grant has made a similar point.

Mature forests are periodically being set back to earlier, transitional stages by logging, forest fires and hurricanes, Sargent theorizes, possibly providing the 'recurring necessity' for melanism that Kettlewell hypothesized. Moths possessing a gene capable of responding to the changed environment by turning black might have survived better than moths that lacked this switch gene mechanism. Such moths might also have an advantage under polluted conditions. The inducing agent, Sargent speculates, could be a foodplant. Soot-covered leaves eaten by caterpillars might turn them black, or other chemical changes in their hostplants, resulting from changes in temperature or solar radiation or some other environmental alteration, might affect their phenotypes. The melanism he induced in *Panthea pallescens* may provide a clue, he thinks. Young white pine trees, with relatively high percentages of new foliage, have very dark bark, he notes, while old white pine trees, with relatively low percentages of new foliage, have very pale bark. Thus the induction may be adaptive. If the caterpillars feeding on new-growth needles turn into melanic moths they will likely be cryptic against the trees of their environment; the same would be true for the light-coloured moths on the light trunks of older trees.[222]

Work by a number of researchers indicates that epigenetic changes – changes in genetic expression that are *not* linked to alterations in DNA sequences – may on occasion be passed on to offspring, in apparent violation of Mendelian doctrine. An article in *Science*, 7 April 2000, dared to suggest that Lamarck might have been 'just a little bit right'.[223] Geneticists at the John Innes Centre in Norwich, England, reported in *Nature* in 1999 that a mutant form of the toadflax plant resulted from an 'epimutation' in which a gene was not expressed. The gene state and the altered flower characteristic were inherited by subsequent generations of the plants. There are other examples of heritable epigenetic changes in other species.[224]

'We didn't use to think that was possible,' says Sargent. 'Kettlewell had great fun at Heslop Harrison's expense when he claimed exactly that. But in cases like forest fire, where it might pay to remain melanic for several years after the fire, could there be some way to do this? There are many scientists who are coming around to the possibility of induced genetic change.'

Lincoln Brower calls Sargent's theory a 'viable hypothesis' that is not Lamarckian. 'I am a certified adaptationist, but I recognize that there can be other processes alongside natural selection such as genetic assimilation. When chemicals exert critical stress for long enough – for example, decades of acid rain – the genome may in some way respond to it. It's a grey area.'

Sargent himself insists that his induction theory is Darwinian rather than Lamarckian, that there would be a selective advantage to having a gene that can switch on or off in response to a recurring environmental change. 'It's like carrying a tool kit,' he says. A caterpillar grows a warm coat in apparent anticipation of winter; with the changing day length in the spring and autumn a bird deposits fat to be used for its migrations; seventeen-year cicadas respond to a seventeen-year rhythm. Just as many animals have evolved to respond appropriately to an upcoming environmental change, these moths may be genetically attuned to environmental cues not yet fully understood, he suggests.

'Is the capital C allele [which codes for the *carbonaria* colour] disappearing,' he wonders aloud, 'or is the capital C allele still there and not being expressed?'

If this question ever gets answered, it will probably not be by Sargent, who often exhibits a *belle indifférence* to the possibility of finding proof for his own induction hypothesis: 'It's just one possibility. It is a better explanation of certain aspects of melanism than the story Kettlewell tells. If there are genes that just have to be turned on it could happen very rapidly.' He has been encouraged to pursue his *Panthea* breeding experiments with greater numbers of moths, but he seems to have run out of steam. Perhaps it is too late in the day. For whatever reason, he seems content to plunk his alternative hypotheses down on the table, perhaps in the hope that a paradigm-smashing young Turk will come along one day and take them to the next level.

'I understand their anger,' he says of his critics. 'I'm very polemical; I'm *trying* to pick a fight. But it won't do any good if we just keep fighting. I don't care if the classical story is right. If it is, fine. But it shouldn't be so sacred.'

It is sacred, however, especially in Darwin's native land, where peppered moth workers naturally view themselves as the stewards of the field. It was one thing for one of their own to admit quietly, at a Linnean Society conference whose pronouncements would be read by a select few, to a few past errors or a few gaps in knowledge, but for renegades from across the Atlantic to start whacking away at the entire edifice was another. The British peppered moth workers had spent decades soberly gathering data on morph frequencies, plotting clines, surveying lichen status in various areas, and elaborating computer models, plugging in values for migration, 'non-visual advantage', changes in fitness with frequency, and so on, and now they were appalled to see their work reduced to near-caricature by a vulgar and uncomprehending popular press.

The *Daily Telegraph* article was the last straw for Laurence Cook, who decided to do something to rehabilitate the image of industrial melanism. 'From being treated as a vivid demonstration of natural selection . . . and good field experimentation, the work concerned has come to be viewed with suspicion,' he lamented. He reanalysed all the data from the major British predation experiments from Kettlewell (1953, 1955) up to Howlett and Majerus (1987) and concluded that overall they showed 'consistent patterns, despite the doubts which have been expressed'. Reviewing the continuous data on peppered moth frequencies in the Wirral from 1959 to the present, gathered by the Clarkes and Bruce Grant, for example, he found some correlation between the fitness of typicals and their frequency that was improved by adding a few assumptions such as the presence of immigrant moths. Cook concluded: 'Considering the variety of techniques used, the lack of control of predation level and the fact that selection exerted during a trial does not estimate the amount experienced in a generation, these experiments seem remarkably consistent.'[225] In short, the various predation experiments, done by different people, with different designs, were all saying the same thing – if you rise above the particulars to the statistical equivalent of an aerial view.

This 2000 paper was the British establishment's riposte to Sargent, Coyne, and the rest of the rebels, and it was greeted with relief and enthusiasm by many. All of Cook's points are well taken: in the real world fitnesses *would* be in flux, not static; immigrants from other areas *would* change the counts; one phenotype might possess unspecified 'non-visual' advantages, and so on. Naturally, it would be naïve to expect a perfect fit between fitness (measured as survival rates in a predation experiment) and frequency at any given time, since the first is a one-time snapshot and the latter an ongoing situation. And, as Cook points out, 'visual selection could be affected by preferential settling behaviour,' and we still don't know whether this happens or not. However, if we need to keep adding refinements and qualifications to make the data work out, where does it all end? Do the data really prove anything at all?

If any anomaly can be rationalized away with a new set of ad hoc assumptions, doesn't this end up rendering the theory of industrial melanism virtually immune to refutation? 'For what good is a theory that is guaranteed by its own internal logical structure to agree with all conceivable observations, irrespective of the real structure of the world?' Richard Lewontin had once asked of the theory of natural selection. 'If scientists are going to use logically unbeatable theories about the world, they might as well give up natural science and take up religion.'[226]

To Sargent, Cook's paper was just one more example of 'special pleading', a fancy way of equivocating and saying, in effect: 'If you look at this, if you don't look at that, it's still a pretty good story.' The British backlash, he complained, also ignored the studies on the North American continent that raised legitimate questions about the classical 'story' of dark backgrounds, lichens, air pollution, and so on. Melanics are equally common in Maine, southern Canada, Pittsburgh, and around New York City. Yet in Sargent's words, 'the British act like no one in North America knew what they were doing.'

According to Karl Popper, a scientific hypothesis should generate predictions that are capable of being falsified, and in Sargent's view the North American data falsify the classical industrial melanism hypothesis. This hypothesis predicts a strong positive correlation between industry (air pollution, darkened backgrounds) and the incidence of melanism. 'But this was not true,' Sargent points out, 'in Denis Owen's original surveys — which showed the same extent of melanism wherever sampled, whether city or rural area — and hasn't been found by anyone since.'

Nature is messy, awash with variables; experiments conducted in a patch of woodland, with its multiply interacting plants and animals, must wrestle with a level of complexity unknown to, for example, experiments with electrons, which have no moving parts. 'Unfortunately, a predation experiment in which differential response is easily demonstrated is likely to be over-simplistic,' Laurence Cook concedes, 'while if it were more realistic (inaccessible sites, a low frequency of exposure to each potential predator and so forth) it would be extremely difficult to conduct.'[227] So must we always look under the lamp-post because the light is better?

Although there have been dozens of experiments since Kettlewell's, no one yet has tried to redo Kettlewell's original experiments *in toto*, using proper controls and statistical techniques as well as live moths. (While Kettlewell used live moths, all of his successors, except for Majerus, have used dead ones, for reasons of convenience.) Knowing what is now known about resting sites, a good experiment might release the moths at the right time of day, just before dawn, and allow them to fly to their own resting sites. 'Kettlewell said you couldn't do it because the moths would just fly to the light trap, but you could just run the light trap on alternate nights or something,' Bruce Grant explains. If one were to try to catch birds in the act of eating moths, as Kettlewell did, one would aim to view this event up near the canopies where moths normally rest.

Not that it would be easy. John Endler once followed a hundred moths to find where they go. 'It's really hard because they go too fast and too high, but I found thirty of them.' The moths he followed were not peppered moths, but he believes someone *should* follow evolutionary biology's favourite icon to find our precisely where it goes during the day. 'If you're talking about crypsis, is it habitat-specific? Tree-specific? You need to know if they only land on certain places or if it's random. You have to know, in the area where they are resting, if the trees are darker or lighter than average – as seen by *birds*. It has to be done properly.'

Michael Majerus has retirement fantasies of following peppered moths to find out where they rest in the wild. He might try to insert a minute radio transmitter in a collar small enough for a moth to wear, let the moths fly all night and settle at dawn the next morning and find them. Or perhaps he'd genetically tag the moths with a chemical marker – green for *carbonaria*, say, and red for typical – then obtain a permit to kill a few birds and see which moths ended up in their guts. (Times have changed; Kettlewell did not have to obtain a permit to shoot gulls and look for moths in their intestinal tracts.) He doubts, however, that he would ever get funding. 'The research councils think they already know the answer, they think it's old hat.'

Endler's landmark 1986 book *Natural Selection in the Wild*, which called for an end to 'quick and dirty studies of natural selection', is widely considered a bible of experimental design. Because his experiments with fish are deemed among the most beautiful demonstrations of natural selection,[228] I wanted to hear what he thought of the original industrial melanism experiments. Kettlewell's studies were a period piece, he explained. If they were done today, certain randomization techniques could be used to eliminate bias; a modern Kettlewell would use a number table to assign moths to particular trees, for example. Of course, Kettlewell could not be blamed for lacking those modern techniques.

What if Kettlewell had gone into the field armed with a hypothesis that he and his all-powerful boss wanted to prove? 'Oh, if he wanted to prove a particular hypothesis, if he was deeply invested in it, that's very bad. You don't want to prove something. You carefully set up your experiment so that you have the right null hypothesis.'

In physics, experimenters do not work for the theorists and have no stake in their theories. When Ernest Rutherford discovered the nucleus of the atom he destroyed two hundred years of classical physics. What he found was not supposed to be there according to the reigning theories of the time, but Rutherford, as director of Cambridge's Cavendish Laboratory, was not beholden to any theorist. In the case of industrial melanism, if we take E.B. Ford as the 'theorist' and Bernard Kettlewell as the 'experimenter' we find the experimenter working for the theorist and under considerable pressure to produce the right answers. That Bernard felt an almost life-and-death urgency about his results was transparent; that E.B. Ford wanted his theories confirmed experimentally was equally obvious.

There are many things we can never know about those long-ago experiments in Dorset and Birmingham, since Kettlewell's field notes have never been found (a circumstance that Bruce Grant calls 'very peculiar'). E.B. Ford, for his part, destroyed all his papers, as well as his breeding material in certain pivotal experiments. Another celebrated experiment of Ford's was possibly fraudulent, certainly irregular. The best that can be said of Kettlewell's studies of dominance and of background choice is that they have not been replicated by a great number of scientists. 'It doesn't happen,' says Bruce Grant, of Kettlewell's dominance breakdown/buildup studies. 'David West tried it. Cyril Clarke tried it. I tried it. Everybody tried it. No one gets it.' As for the background-matching experiments, Mikkola, Grant and Sargent, among others, repeated what Kettlewell did and got results contrary to his. 'I am careful not to call Kettlewell a fraud,' says Bruce Grant after a discreet pause. 'He was just a very careless scientist.'

Thomas Kuhn observed in *The Structure of Scientific Revolutions* – a book that, like *On the Origin of Species*, everybody quotes but few have read – that scientists work within the confines of certain doctrines and customs, easily becoming 'locked in' to a body of acceptable thought, a 'paradigm', in short. 'No part of the aim of normal science,' Kuhn wrote, 'is to call forth new sorts of phenomena; indeed those that will not fit in the box are often not seen at all. Nor do scientists normally aim to invent new theories, and they are often intolerant of those invented by others.'

The biologist Adrian Wenner believes that 'entire careers and the self-images of innumerable scientists may appear to be undermined [by the fall of a paradigm]. The longer a hypothesis is considered "established fact," the more difficult it becomes to accept another test of it.'[229] Wenner should know. He challenged a major icon of biology, the waggle-dance of the honeybee, the life's work of the great ethologist Karl von Frisch, who shared a Nobel Prize with Niko Tinbergen. Unprepared for the vehemence of the reaction, Wenner walked confidently to the blackboard at the Salk Institute one day and drew sketches and diagrams illustrating how von Frisch's results, because of a lack of adequate controls, could equally fit a different model. 'The audience reaction immediately turned from what might best be termed one of euphoria to one of intense hostility . . .' he recalled. 'The reaction of [an eminent psychologist] was fairly typical; he shouted: "What's the matter? Don't you believe anything unless you have done it yourself?"' Over the years Wenner was excluded, attacked, reprimanded, shunned and stigmatized by the community of his peers. He and Sargent once spoke of collaborating on a book about the experience of being on the wrong side of a paradigm.[230]

The paradigm underlying Kettlewell's work was a web of beliefs about natural selection. Certain presumptions were clear even in his first *Heredity* paper, in which Kettlewell declares just after reflecting on the rarity of seeing birds eating moths off tree trunks: 'Nevertheless, the effective concealing patterns found in great numbers of these insects (those affected by melanism and others) seem explicable only on the assumption that predators hunting by sight are of serious danger to them.' In other words, Kettlewell took it for granted that the wing patterns *must* have the function of protecting the moths from predators; they must have adaptive value.

'If Kettlewell hadn't been so convinced of the truth of bird predation affecting peppered moth evolution,' Craig Holdrege writes, 'he might have left more room for alternative explanations . . . When scientists have, as Lynn Margulis puts it in *Slanted Truths*, "an uncritical acceptance of the mesmerizing concept of adaptation," there is real danger of seeing what one believes.'

Science proceeds via hypotheses, and in an ideal world scientists attempt to falsify those hypotheses. Becoming overly attached to a particular hypothesis tends to stall the process, however, and it is well known that some entrenched scientific beliefs, such as static continents, have proved difficult to eradicate. H.B.D. Kettlewell advanced his hypothesis and tested it, though perhaps not as rigorously as one might have wished, and it seemed fine for a time. Today new evidence, a rigorous re-examination of old evidence, and further tests of the hypothesis (or, really, predictions of the hypothesis) have suggested that it may be incorrect. The next step, in Sargent's view, is to try harder to falsify the original hypothesis, to devise alternative hypotheses, and to construct tests that would clearly eliminate, or falsify, one or another of these hypotheses.

When the Nobelist Sir Peter Medawar famously declared: 'The science paper is a fraud,' he was referring to the fact that science doesn't always proceed according to Popper's ideal of the hypothetico-deductive method, but he may also have been thinking of the human factors (ambition, jealousy, rush to publication, megalomania, behind-the-scenes manoeuvrings, even sex, illness, desperation) that are sanitized out of the journals. Underlying the traditional science paper, as Karl Sabbagh observes in *A Rum Affair*, the tale of Heslop Harrison, 'are layers of unreported facts – facts that could never be reported without legal advice – about U's incompetence, V's marriage break-up, W's burning desire for a Nobel prize, X's dislike of experiment, Y's irrational belief in a particular theory against all evidence, Z's need to please his or her boss'.

We could fill in the blanks: Bernard's desperate need to please E.B. Ford, Ford's fanatical belief in the power of adaptation, everybody's craving for merit badges (an FRS for Bernard, a knighthood for Henry perhaps). Then there were Bernard's family problems, his insecure finances, his craving for recognition at Oxford, the 'turns' he suffered during his experiment, his exhaustion, sleep deprivation and mood swings, even the emotional aridity of his childhood. We might add Henry's desire for revenge on his enemies, his idolatry of Sir Ronald Fisher, the intrigues and turf wars within the zoology department at Oxford, his investment in his experiments with *Maniola jurtina* and *Panaxia*, his grandiosity, even his discomfort with females. Traditionally, these human factors are deemed irrelevant to the scientific process, as if scientists were a race of hard-wired robots.

As I looked into the world of moth-hunters, I sensed I was in the presence of a powerful compulsion I did not understand. There seemed to be something primal about this intimate communion with creatures so phylogenetically distant from ourselves. Lepidoptery was obviously addictive, a fever of the mind contracted, mainly by boys, on the verge of adolescence, and the consoling powers of moth-hunting on the lonely, the unloved and the bereaved seemed self-evident in the personal stories of the scientists I learned about. Lonely, friendless Edmund Ford. Stifled Bernard Kettlewell. And wounded Ted Sargent.

The subtext of loss and bereavement in Sargent's writing

seemed unusual in a biologist. Even in his book on underwing moths, the tragic roster of *Catocala* names – the Forsaken, the Dejected, et cetera – seemed to mirror some secret sorrow in the author's soul, as did the quote on its frontispiece: 'Every association of moths is with night and mystery and death.' I had noticed an air of stoic resignation about him, a strain of melancholy, or perhaps something like self-negation. Clearly, he seemed to possess a rare gift for passing unnoticed, and yet he was a firebrand. At first I chalked this up to his professional disappointments; later I learned that it went deeper.

Ted Sargent was a year old when his mother died giving birth to twin girls, leaving three infants in diapers and an eight-year-old. Reeling from the scope of the tragedy, his optometrist father absented himself psychologically – 'He wouldn't even eat with us' – and married a woman who was like the archetypal stepmother in a fairytale. Superficial, strangely devoid of affect, obsessively prim and puritanical, she beat all the children whenever they acted 'uppity'. She tossed out Ted's moth collections whenever she came across them, and sent him to school with a virtual 'kick me' sign on his back – dressed in a vest, knickers, and patent leather shoes, with steel-rim glasses and hair parted in the middle. 'I got killed every day. I had to walk four miles out of my way to get home without getting beaten up.' Only Protestant friends were allowed in the house. Ted's Catholic friends were permitted to come as far as the piazza, Jewish friends only up to the gate.

It doesn't take much of a psychologist to see how Ted Sargent's childhood might have bequeathed him a lifelong psychic invisibility such that, like his underwing moths, he blends in with the scenery. In childhood he would have survived by escaping notice, cultivating a quiet, unoffending exterior and taking refuge in the outdoors and its wild creatures.

'I spent a lot of time outside watching birds. They were more interesting to me than people, and certainly less cruel.' Birds and moths had saved him somehow, as they had saved Bernard Kettlewell and E.B. Ford, and perhaps they also offered a valuable lesson in protective self-camouflage.

In addition to moths and birds, Sargent seems to take refuge in more than the usual complement of hobbies. I learned that he wrote poetry; that he was a rare book collector; that he was working on a study of birdsong in nineteenth-century poetry; that he was investigating the phytochemical properties of herbs used in witchcraft and planned to write a book about *that*; that he had once owned a store on Cape Cod, selling antiques, rocks and minerals. Like his moths, he had a nocturnal life for a while, playing the Dobro (a steel guitar with an aluminium resonator, played while held flat) in a blue-grass band in smoky local bars. 'I wanted to know what that world was like,' he says. A politically self-defeating biology professor with 'Wiccan' friends and obscure poetical interests might strike many colleagues as an oddball, and, characteristically, Sargent did little to deflect this impression: 'Maybe they're right. Maybe I am a dilettante.' As if to reinforce the point, he is currently at work on a biography of a precocious Massachusetts child-poet of the nineteenth century named Elaine Goodale.

'At one time Ted was spending his summers and autumn weekends at his store on the Cape,' says his friend John Moner, a professor emeritus of cell biology at U Mass. 'A lot of people in the department didn't approve of that. They thought he should be doing research. But Ted is the most meticulous, careful researcher. When he gets interested in something – if it's rocks and minerals, or herbs, or melanism – he pursues it with tremendous intensity.'

Like his nemesis, Kettlewell, but for different reasons, he seems to have been temperamentally out of tune with academic life. 'I just can't stand it,' he sighs. 'I can't even sit through dinner sometimes. They're going on about what journals to publish in, or the people that you should know, or the right places to go to collect. If you're verbally facile and reasonably bright you can fake it in science and be famous, at least at the local level.'

It may or may not be evidence of a shifting paradigm that Sargent's 1976 out-of-print book on underwing moths, in which his renegade industrial melanism theories make a cameo appearance, is suddenly selling on some internet used-book sites. At the same time, however, his latest paper on background choices of birch-feeding moths continues to be rejected by the scientific journals. In his experiment, he tells me, both melanics and typicals selected white backgrounds and both rested on birch trees. He believed, and still believes, that this has important implications for the industrial melanism story, but when the journals sent his manuscript out for review it was repeatedly savaged by one hostile reviewer whom Sargent came to recognize. 'It's always the same typewriter.'

One evening at a dinner party I mentioned the peppered moth, and our host, a high school science teacher, said: 'Oh, have you seen the kits?' He procured one for me. For an odd, disassociative moment, it was possible to imagine real peppered moths fluttering out and settling on a strip of Dorset bark, even that one might catch the ghostly cadences of Bernard Kettlewell's or E.B. Ford's Oxford accents. Of course, there were only polystyrene bags full of little paper moths, light and dark, and several paper 'bark simulations' on which to stick them. Following the instructions, groups of students are guided to enact the roles of Kettlewell, the birds, and the operation of natural selection. 'From an arm's length away from the tray,

Participant C removes, one at a time, as many moths as possible within 4 seconds and then looks away from the background. In the next step, Participant B replaces each black moth removed with a white moth. White moths removed by Participant C are replaced with black moths . . .'

I tried communicating with Science Kit & Boreal Laboratories in Tonawanda, New York, which makes the kits. I wanted to know their scientific basis and tried to discover whom the kit makers had consulted. Did evolutionary biologists review the kit's instructions to make sure that shuffling these little cardboard moths would result in a valid experiment?

My questions went unanswered as I was shunted from one person to the next, each one stunned anew by my queries. (Obviously, no curious science teacher had ever called this number.) The bottom line was that there are no 'labs' at Boreal Labs. It's a warehouse that buys the stuff from subcontractors. No one there knew anything about the kit's origins. You might as well ask about the origin of life.

In a sense, real peppered moths are treated as if they were the little cardboard tokens in the kit. Kettlewell's experiments are often defended on the grounds that they are a superb pedagogical tool. Even if it is a little bit wrong, some say, we should stick with the basic story because a more complex one would confuse students.

'If you're a teacher you look for a good story,' observes one teacher of the history of science at the State University of New York. 'The peppered moth is like George Washington chopping down the cherry tree. It's a damn good story.'

It *is* a damned good story, a narrative so satisfying, so seductive, that no one can bear to let it go. But a good story alone is no substitute for truth. Jerry Coyne 'probably won't' teach the textbook story to his students until the 'agent of selection' is identified. Douglas Futuyma agrees that 'it should not be taught in the simplified form it has been.' Richard Harrison, an evolutionary biologist at Cornell, is emphatic: 'We should no longer use the example in textbooks.' Craig Holdrege, the young biology teacher who, dismayed by the flaws in the textbook example, quit his teaching job rather than repeat the standard tale, describes it as 'filled with half-truths. This is not because teachers and writers are intentionally lying, or hiding and bending facts, but because the example is only brought to prove a point.'[231]

Holdrege now teaches biology at the Hawthorne Valley School in Ghent, New York, where his students do not learn that the peppered moth is a gemlike example of evolution in action. They linger over industrial melanism much longer than the time allotted in public high school biology classes. They ask questions about the peppered moth's natural history: Where does it rest during the day? What are its natural predators? How far can it fly? How long do the moths live? They do not use peppered moth kits (Holdrege calls them 'indoctrination kits': they make him 'cringe') and they are taught that the idea of natural selection need not be 'like a pair of spectacles that one doesn't remove any more'.

By coinciding with the era of Charles Darwin, industrial melanism gained an immortal place in science. By the time a baby born today reaches college, black peppered moths may be gone from the planet, yet their pictures, posed on lichened and unlichened trunks, may live on in the textbooks, Dorian-Gray-like, illustrating 'Darwin's missing evidence'. Students then, like students now, will study these pictures as a modern parable, little suspecting the ambitions, rivalries, *idées fixes*, power plays, disillusionments and broken dreams that they conceal.

There seems to be an irreducible mystery at the heart of

the industrial melanism story, one that is still unsolved. These jet-black moths arrived in the middle of the century before last, surprising a calico maker/lepidopterist in his tiny Manchester garden; the moths thrived, grew in number, and stayed for over a hundred years; now they are swiftly disappearing. Majerus wants to track the *carbonaria* form in Britain until it disappears, as he is 'reasonably confident' it will. 'It had been ninety per cent in Cambridgeshire; now it's down to ten per cent. I think it will get to less than one per cent in 2019. I'd love to see whether that's true.' Even to Majerus, who has spent much of his life studying them, aspects of the peppered moths remain veiled.

When Bernard Kettlewell published his landmark article, 'Darwin's Missing Evidence', in *Scientific American* in 1959, the ecological message was starkly clear. 'It happens that Darwin's lifetime coincided with the first great man-made change of environment on earth,' he wrote. 'Ever since the Industrial Revolution commenced . . . large areas of the earth's surface have been contaminated by an insidious and largely unrecognized fallout of smoke particles.' The mutant black moths were like canaries in a coal mine, the visible marks of an environmental catastrophe. Today, as the melanic moths disappear, and sulphur dioxide and particulate levels decline, it is easy to assume that we know the ending of this story, and that it is a happy one.

But is it? While it is true that there may be fewer particulates in the atmosphere, the overall environmental picture has grown grimmer, if anything, certainly for wildlife. According to J.R. McNeill, the author of *Something New Under the Sun*, species are disappearing at an unprecedented rate, due largely to habitat disruption brought about by the activities of humans, and experts expect '30 to 50 per cent of terrestrial species to disappear in the next century or two'.[232] Every time a highway, or even a small country road, slices through a forest, or a field of wildflowers, an ecological tragedy may ensue. To the tourists breezing through at 70 miles an hour the scenery may look like a wilderness; they do not see that the habitat of the animals has been halved, leaving many with no space in which to escape from edge predators. Some populations become marooned in populational 'islands', cut off from others of the species and doomed to extinction within a few generations. Rainforest disruption in distant states or countries has destroyed the wintering sites of many birds, which then disappear from our midst. Habitat disruption and forest fragmentation, air pollution, pollution of the waterways by toxic chemicals, roads, automobiles, are perpetrating what has been described as the last Great Extinction.

Most of us, with cell phones glued to our ears, hardly notice, but those who are familiar with birds know that the world used to be richer in song. Every spring, on Memorial Day, for the last thirty years, Ted Sargent has sat on his porch in Leverett, counted the birds and made a list, and his list is growing shorter. He hasn't seen an oriole or house wren or hermit thrush for years. The haunting midnight cries of the barred owl are no longer heard around his house, for the mice, voles and birds they used to eat are gone. The regal fritillary butterflies he collected as a boy in his butterfly net have vanished. Gone too are the ring-necked snakes, spotted salamanders, box turtles and wood turtles his children used to catch twenty years ago. 'We're losing everything,' he tells me, 'our birds, our butterflies, our moths, our mammals, our reptiles.'

In an odd case of symmetry, the people who study birds, butterflies, fish, mice, snakes and wolves in the field – the 'whole-animal' biologists – are vanishing too, supplanted by the newer caste of molecular biologists, who rarely venture outdoors. The biology department of the University of Massachusetts, for instance, hired its last animal ecologist twenty-five years ago, and has not replaced most of the last fifteen to twenty retiring field biologists, including Sargent. Any lepidopterist who is hired today is fairly certain to be a molecular systematist, who grinds up the insects and studies their DNA. One overhears them announcing proudly: 'We're not only doing dragonflies; we can do ladybugs and bats.' 'It's true here too,' says Austin Platt, of the University of Maryland Baltimore County. 'Our newest ornithologist is doing DNA studies of orioles and ravens.'

Whole-animal biologists I interviewed worry that entire worlds of knowledge are disappearing with the academic positions. The old-time lepidopterists, who know where to find a species, what it feeds on, what its flight season is, are dying out and no one is replacing them. There is no longer a first-rate specialist in noctuids in the USA, at any university or museum, Sargent tells me. Because most of today's grant money is in the molecular work, there is little demand for someone who can train students to conduct life-history studies of butterflies and moths.

The extinction of realms of natural history knowledge, together with the wildlife, is surely a part of this story. 'I think,' muses Sargent, 'that's one reason that extensive sampling for melanics, and critical experiments involving these moths, are not being done today. You get a big trap full of a thousand moths and you have to find the ones you want. Very few know how to do that any more.' Bernard Kettlewell was one of this vanishing tribe, a 'fine specimen of an English type rapidly approaching extinction', in the words of Arthur M. Shapiro, who did fieldwork with him in the summer of 1969. Although 'his experimental designs were flawed,' Shapiro, now a population geneticist at the University of California at Davis, observed recently, '. . . the important thing is that he did it. Nobody else did. And to our shame, no one did anything similar in America, despite ample opportunities.'[233]

While researching this book, I was repeatedly asked by biologists what 'my agenda' was, or whether I 'believed in evolution'. Many scientists said they had been burned by creationists masquerading as reporters, and their names had turned up in embarrassing venues, bracketed by Bible quotes. When I said I wanted to discuss the peppered moth, a few reacted as nervously as if I had just said I was calling from the Internal Revenue Service. There is a reason for this paranoia. By the time you read these words, passages from this book will be strewn across dozens of creationist tracts and propagated to the far corners of the internet, no doubt creating the impression that the book is a creationist manifesto.

As anyone who has been reading the newspaper knows, there are two types of creationism. The old-fashioned Biblical-literalist creationists tend to believe that the Earth was created at some recent date (and that the fossils are therefore lies, perhaps sown by the devil) and that life on Earth unfolded more or less as specified in Genesis. The scientifically sophisticated 'soft creationists', in contrast, concede that the Earth is 4.5 billion years old, that fossils are real, and that 'descent with modification' – that is, evolution – did occur. They draw the line at natural selection, however, insisting that it cannot account for the diversity and complexity of life, and that some sort of 'intelligent design' must be guiding the process. Intelligent design (known in the trade as 'ID') may take the form of God, or something more like an *élan vital*, an inner purpose, or even something that originated in a meteorite from outer space.

Since getting wind of its weaknesses, the creationists have turned the peppered moth affair into a *casus belli* – which is hardly surprising, given the number of times these moths have been marshalled as unassailable proof of evolution in the past. 'One of the arguments of the creationists,' Isaac Asimov wrote in 1984, in a passage repeated in one form or another by countless other science writers, 'is that no one has ever seen the forces of evolution at work. That would seem the most nearly irrefutable of their arguments, and yet it, too, is wrong. In fact, if any confirmation of Darwinism were needed, it has turned up in examples of natural selection that have taken place before our eyes ... A notable example occurred in Darwin's native land.'[234] Whereupon Asimov proceeded to tell the peppered moth story.

Nowadays on creationist websites Coyne, Majerus and Sargent, none of whom are creationists, are invoked like patron saints in articles entitled 'Let's mothball the peppered myth' and 'The Piltdown Moth', and the best sound-bites, such as Coyne's quip about Santa Claus, are cloned repeatedly. 'Evolutionists' are typically portrayed as conspirators engaged in a vast plot to suppress the truth. In the face of this increasingly vocal, well-organized and sophisticated creationist threat, evolutionary biologists are circling their wagons. Quoted in a front-page article of *The New York Times* in April 2001, Jerry Coyne referred to the intelligent design argument as 'devilishly clever. It has an appeal to intellectuals who don't know anything about evolutionary biology, first of all because the proponents have PhD's and, second of all, because it's not written in the sort of populist, folksy, anti-intellectual style.' Like Coyne, most mainstream evolutionists view ID as little more than lace-curtain creationism and try to distance themselves from it as far as possible.

Perhaps the creationist attack would not appear so menacing

if scientists had not been happily 'coasting' on Kettlewell's experiments for years, Arthur Shapiro suggests. 'Creationists are sometimes outright intellectually dishonest, but perhaps their belief that natural selection rises or falls on the peppered moth is at least a little justified by how proud most of us were of that tale.'[235] The peppered moth has been transformed into a potent symbol of the triumph of Darwinism, an amulet against the forces of darkness and superstition. I was struck by how Michael Majerus, in the course of a long transatlantic phone conversation, riffed seamlessly from bird predation to a sweeping endorsement of evolution: '. . . and that this funny thing, natural selection, does exist, and therefore evolution does exist. And if you're suddenly asking what caused life to be the way it is — was it special creation or biological evolution? — there is only one answer you can come to if you look at the peppered moth.' For creationists, conquering the peppered moth is like capturing the enemy's flag. Anti-evolutionist parent committees critiquing new biology texts seem terrifyingly well informed about lichens and UV vision. In the current cultural cold war, many biologists fear that anything less than a united front will play into *their* hands. Scientists who express too many doubts run the risk of being considered traitors or turncoats.

'We're in conflict with fundamental religious beliefs,' assents Cornell's Richard Harrison. 'But are we fudging on data to bolster Darwin's theory of evolution?'

By removing the supernatural order, the Darwinian revolution came as a profound shock to the collective psyche, jolting Victorians out of a placid natural world of nested hierarchical life-forms, each in its appointed place, into a sprawling jungle of perpetual conflict and bloodshed, in which God, if He existed, seemed to be an impotent bystander. The 'survival of the fittest' became translated into a widespread social Darwinism that the robber barons of the Gilded Age used to justify the worst kind of capitalist cut-throatedness; it was no accident that Andrew Carnegie became an ardent endower of naturalistic endeavours. As late as 1921 George Bernard Shaw was still railing against Darwinism for sucking the meaning out of the universe. Darwin himself worried that a natural world ruled by natural selection instead of God might be empty of meaning and purpose, and the traces of teleology in his writing (notably, the unforgettable image of Natural Selection 'daily and hourly scrutinizing . . . the slightest variations . . . silently and insensibly working . . . at the improvement of each organic being') are a psychological hangover from the old natural theology he grew up on. Over the years, natural selection began to fill the ontological vacuum left by the old-fashioned creator God, emerging as a kind of cosmic force that 'acted' on things, 'moulding' organisms into ever more perfect creatures. Everything on Earth, from the stamen of a tulip to the toenail of a tree sloth, told a story of adaptation. It was the new religion.

Many of these adaptive stories sounded plausible, but they were often wholly conjectural, as evolutionary biologists themselves realized. Where was the proof? Where were the data? Natural selection is a principle that seems self-evident, yet turns out to be very, very hard to catch in the act, nature being full of so many things. But Bernard Kettlewell appeared to do it. He drew a circle around a swatch of English woodland and captured natural selection in black and white: a two-to-one selective advantage in one place, a three-to-one advantage in the other (or two-to-one, depending on which figures you accept). Or so it seemed for a while.

Reassessing Kettlewell's experiments, and those that followed, we may conclude that further work needs to be done to pin down the workings of natural selection in this case. But this is emphatically not a victory for creationism or for anybody's God in particular. A few years ago, during the uproar over Gould and Eldredge's theory of punctuated equilibrium, creationists insisted that if Darwin was wrong about the tempo of evolution he was mistaken about everything. Now the peppered moth's troubles are being seized upon as evidence against Darwin. But the fact that 'Darwin's missing evidence' is imperfect does not disprove the theory of evolution. Perhaps the peppered moths, like overburdened donkeys of some Greek islands, have simply had to carry too great a load. It is reasonable to assume that natural selection operates in the evolution of the peppered moth. Undoubtedly, it is an absorbing story, and a vitally important one, involving many factors besides bird predation as yet unidentified. It is a juicy puzzle that remains to be solved.

The periods in science when things start to fall apart are sometimes the most interesting times. When the anomalies multiply, and a theory requires propping up with more and more ad hoc provisos – such as the epicyles that had to be added to Ptolemaic theory to accommodate the observed data on planetary motion – it may mean that a paradigm is coming unglued, in Kuhnian terms. When this happens, the philosopher of science Michael Ruse writes in *Mystery of Mysteries*,[236] at first 'everyone works more frenetically to shore things up . . . Then, however, if one is lucky, someone – usually someone young or new to the field . . . puts forward a new paradigm . . .' The scientific community switches its allegiance, and 'normal science' resumes. A key part of this process, Ruse notes, is that the textbooks get rewritten.

The tale of the peppered moth has centred on two protagonists, united and divided by a common obsession. One was British, the other American; one an amateur, the other a well-trained biologist; one a loud, excitable extrovert who instantly took over any room he entered, the other a shy, bookish, meticulous introvert with a talent for being overlooked. They met only once, and did not hit it off. Their encounters in print were even less amicable. They were professional enemies whose fates became intertwined with the mystery of a particular moth, which would end up defining their careers and breaking both their hearts. Both men had transcended unhappy childhoods through the entranced observation of insects. Both were moth men who had earned their spurs in the old-fashioned natural-history world of killing bottles and larval foodplants, and who found their deepest satisfaction outdoors rather than in lecture halls or departmental meetings. Both became marginalized in their academic milieux and chafed at the servitude imposed by bosses and overseers.

If the story were told by the accomplished amateur lepidopterist Vladimir Nabokov, Kettlewell and Sargent might be revealed as doppelgängers, mythic doubles, like *Lolita*'s antihero Humbert Humbert and Clare Quilty, his pursuer and nemesis. Small, seemingly meaningless resemblances, like the fact that Kettlewell and Sargent both had sons named David, would have been the clues to a hidden pattern, and in Nabokovian hands the moths themselves, black and white versions of the same organism, locked in perpetual Darwinian rivalry, might have been doppelgängers, too. Admittedly, Nabokov seems not to have had much truck with moths, or even particularly with evolution. However, an undercurrent of many of his novels is the peculiar intimacy, even empathy, that can grow up between enemies who are secret sharers, between the pursuer and his quarry. In this most intense of bonds, every slight is magnified, every word or gesture becomes imbued with meaning, as in the dialogue of lovers.

To Sargent, Bernard Kettlewell became the incarnation of the forces that damaged his career; for Kettlewell, Sargent seems to have been a hidden irritant and something of a guilty secret. He thought about him more than he let on.

'So you have met Theodore Sargent,' he wrote to an acquaintance in 1970. 'I shall be interested to hear what you think of him. I have a feeling that I am "persona non grata" as I just cannot agree with his designs of experiments . . . work he was done on background choice . . . and have said so. He uses containers (which incidentally are not cylinders) of very small cubic capacity and I have no doubt that many of the species incarcerated there would act in an entirely abnormal way . . .'[237] Whenever Sargent's name came up, he would repeat some version of the same defensive critique. In 1974 a mutual friend, a well-known Connecticut lepidopterist, wrote to Kettlewell to say that, by the way, Sargent was a 'bit indignant over something you put in, or left out of, the melanism book'. Kettlewell wrote back: 'I am sorry I have upset Sargent; I had learnt this already. There were two reasons for this and I think both are valid. Firstly I left out one of his pieces of research (quite accidentally) in which he had actually worked on a dimorphic species. Secondly, I left his name out of "the Author and Contributors" list – for this I am indeed sorry.'[238] Hearing about this letter, a wounded look passed over Sargent's face. 'Gee, I wish I had known he was sorry. He never said that to me.'

Perhaps it takes a Clare Quilty to understand Humbert Humbert and his sins; and, for all his criticisms, Sargent respected Kettlewell's field knowhow and understood better than anyone what it must have taken to pull off those historic experiments.

'The whole breeding thing is trial and error,' he reflects. 'You have to feed the larvae, get them to survive the winter, time their emergence, keep them in the refrigerator. There are all these critical things. The wings get distorted if they are too wet or too dry. How did Kettlewell do it all? The assembling traps, the mercury vapour traps, the sleeves . . . You have to give him a lot of credit. The sheer difficulty of it was mind-boggling.'

Notes

1. Jack A. Ward and Howard R. Hetzel. 1980. *Biology: Today and Tomorrow*. St Paul, Minn.: West Publishing Company.

2. Craig Holdrege. Spring 1999. The case of the peppered moth illusion. *Whole Earth*.

3. Jean Henri Fabre. 1916 (1915). *The Life of a Caterpillar*. Tr. by Alexander Teixera de Mattos. New York: Dodd Mead and Co.

4. Ekkehard Friedrich. 1986. *Breeding Butterflies and Moths: A Practical Handbook for British and European Species*. Colchester, Essex, UK: Harley Books.

5. See Theodore S. Sargent. 1976. *Legion of Night: The Underwing Moths*. Amherst, Mass.: University of Massachusetts Press.

6. I am indebted to Michael E.N. Majerus, author of *Melanism: Evolution in Action* (1998, Oxford University Press), for many details of the early history of melanism.

7. This crucial sequence of events was reconstructed by Frank J. Sulloway in 1979 in 'Geographic isolation in Darwin's thinking: the vicissitudes of a crucial idea', *Studies in the History of Biology* 3:23–65.

At first, Darwin thought geographical isolation was necessary; later he modified his views to include the possibility that species could evolve due to specialization within habitats.

8. Charles Darwin. 1958 (1859). *On the Origin of Species*. New York: New American Library.

9. Darwin was not dogmatic, however. While he regarded natural selection as the chief mechanism, he also accorded a role to the inherited effects of use and disuse à la Lamarck.

10. George Gaylord Simpson. 1944. *Tempo and Mode in Evolution.* New York: Columbia University Press.

11. Charles Darwin. *On the Origin of Species.* Op. cit.

12. William Paley. 1836. *Natural Theology*, vol. 1. London: Charles Knight.

13. Ernst Mayr. 1991. *One Long Argument: Charles Darwin and the Genesis of Modern Evolutionary Thought.* Cambridge, Mass.: Harvard University Press.

14. Charles Darwin. *On the Origin of Species.* Op. cit.

15. Steve Jones. 2000 (1999). *Darwin's Ghost: The Origin of Species Updated.* New York: Random House.

16. Michael E.N. Majerus. *Melanism: Evolution in Action.* Op. cit.

17. W.B. Provine. 1971. *The Origins of Theoretical Population Genetics.* Chicago: University of Chicago Press.

18. Joan Fisher Box. 1978. *R.A. Fisher: The Life of a Scientist.* New York: John Wiley and Sons.

19. R.A. Fisher. 1930. *The Genetical Theory of Natural Selection.* Oxford: Clarendon Press.

20. R.C. Punnett. 1915. *Mimicry in Butterflies.* Cambridge: Cambridge University Press.

21. Ernst Mayr and William B. Provine. 1980. *The Evolutionary Synthesis: Perspectives on the Unification of Biology.* Cambridge, Mass. and London: Harvard University Press.

22. Julian Huxley. 1970. *Memories.* New York: Harper and Row.

23. 'Tell me, do you ever read *The Daily Express?*'

'Never,' said Shearwater. 'I have more serious things to do.'

'And what serious thing, may I ask?'

'Well, at the present moment,' said Shearwater, 'I am chiefly preoccupied with the kidneys.'

24. Charles Darwin, letter to George Bentham, in Francis Darwin (ed.). 1911. *The Life and Letters of Charles Darwin, Including an*

Autobiographical Chapter. New York, London: D. Appleton & Co. Vol. 3, p. 25.

25. Ford's reminiscences are quoted in Ernst Mayr and William B. Provine. *The Evolutionary Synthesis*. Op. cit.

26. David A. Jones, professor and former chair of the Department of Botany of the University of Florida, Gainesville, was one of the last two undergraduate students of R.A. Fisher in the Department of Genetics at Cambridge. He did his graduate work at Oxford under E.B. Ford, earning a doctorate in 1963.

27. Miriam Rothschild. 1984. Dedication: Henry Ford and Butterflies. *The Biology of Butterflies*. R.I. Vane-Wright and P.R. Ackery (eds). New York and London: Academic Press.

28. Interview with Dame Miriam Rothschild.

29. H.G. Wells, Julian Huxley and G.P. Wells. 1929. *The Science of Life*. London: Cassell & Co.

30. Ibid.

31. Karl Sabbagh's *A Rum Affair* (Allen Lane: The Penguin Press, 1999) is a delightful chronicle of Heslop Harrison's life, particularly his botanical frauds.

32. Ford especially liked to drop the names of Darwins. Throughout his life he would frequently wax nostalgic about his intimacy with members of the Darwin family, as well as important early Darwinists. 'I am in a slightly odd position – I believe I am the very last of those who could and did share friends with Darwin,' he announced at the beginning of a taped general lecture on genetics.

On another occasion he wrote: 'Among those much older with whom I was *intimate* [his emphasis], I would especially mention Major Leonard Darwin [Charles Darwin's son] and Sir Ray Lankester [a notable Oxford Darwinist]. He was a friend of Darwin's and twice went to stay with him at Down. I could, therefore, and did repeatedly, ask them what Darwin said to them on this or that.' This was part of Ford's reply to a biographical 'Questionnaire Concerning the Evolutionary Synthesis', which he was invited to complete for a 1974 meeting of Synthesis leaders in the United States, organized by Ernst Mayr and William B. Provine. (MS. E.B. Ford papers. Bodleian Library. University of Oxford.)

33. Letter from E.B.F. to H.B.D.K. 1937. E.B. Ford papers. Bodleian Library. University of Oxford.

34. Interview with the late David Kettlewell.

35. R.F. Demuth. 1979. Obituary: Dr H.B.D. Kettlewell. *Entomologist's Record and Journal of Variation* 1/X/79.

36. Ibid.

37. Letters from E.A. Cockayne to H.B.D.K. 25 May 1948 & July 1952. H.B.D. Kettlewell Collection. Bodleian Library. (The Kettlewell papers are owned by Wolfson College, Oxford.)

38. Bryan C. Clarke, FRS, now emeritus professor of genetics at the University of Nottingham, was a graduate student in the Oxford School of Ecological Genetics from 1955 to 1959.

39. Interview with Laurence Cook.

40. Interview with Lincoln Brower. Now research professor of biology at Sweet Briar College and distinguished service professor of zoology, emeritus, at the University of Florida, Brower was a graduate student at the Oxford School of Ecological Genetics in 1957–58, and a postdoc in 1963–64.

41. E.B.F. to H.B.D.K. 27 January 1951. Kettlewell papers. Bodleian Library.

42. E.A. Cockayne to H.B.D.K. 12 April 1951(?). H.B.D. Kettlewell papers. Bodleian Library.

43. I am indebted to the late David Kettlewell for a copy of this colourful and entertaining diary.

44. On this subject Cockayne was characteristically gloomy, chiding H.B.D.K. in September 1952: 'The number of pupae is not bad but not enough for the experiments you propose to carry out and why didn't you put covers over your tubs? It is a pity to take all the trouble to feed larvae to full growth and then let pests eat the pupae. I still think you are much too optimistic about the percentage of returns you will get.' H.B.D. Kettlewell papers. Bodleian Library.

45. H.B.D.K. to W. Bowater. 30 November 1951. H.B.D. Kettlewell papers. Bodleian Library.

46. H.B.D.K. to W. Bowater. Ibid. The letter continues: 'I am therefore thinking of working on a second species of melanic at the same time. It has got to be one that is widely distributed or spreading. What do you think about me working on *bidentata?*' *Gonodontis bidentata*, the scalloped hazel, was another British species with a melanic form, known to be relatively sedentary. Bernard was hedging his bets.

47. John A. Endler. 1986. *Natural Selection in the Wild*. Princeton, NJ: Princeton University Press.

48. In his book *Genetic Polymorphisms* (MIT Press, 1965), reflecting on the apparently trivial features characterizing *Drosophila* mutants, such as the number of bristles, Ford wrote: 'Yet there has never been a one of them that does not affect viability; altering length of life, capacity to survive in unfavourable conditions, male fertility, or the number of eggs laid per unit of time. That is to say, the genes in question, insignificant as are their visible effects, have an important influence upon the physiology of the organism, modifying profoundly the individual as a unit upon which selection operates.'

49. This account comes from Arthur Cain, quoted in W.B. Provine. 1986. *Sewall Wright and Evolutionary Biology*. Chicago: University of Chicago Press.

50. E.B.F. to H.B.D.K. 11 October 1949. H.B.D. Kettlewell papers. Bodleian Library.

51. A.J. Cain and P.M. Sheppard. 1950. Selection in the polymorphic land snail *Cepaea nemoralis* (L.). *Heredity* 4:275–94.

52. A.J. Cain. 1964. The perfection of animals. In J.D. Carthy and C.L. Duddington (eds). *Viewpoints in Biology 3*. London, Washington, DC: Butterworth's.

53. John R.G. Turner. 1991. Stochastic processes in populations. In R.J. Berry, T.J. Crawford, G.M. Hewitt (eds). *Genes in Ecology*: The 33rd Symposium of the British Ecological Society, University of East Anglia. Oxford: Blackwell Scientific Publications.

Turner recently retired from his position as professor of evolutionary biology at the University of Leeds.

54. Not his real name.

55. The aviary experiments are reported in H.B.D. Kettlewell. 1955. Selection experiments on industrial melanism in the Lepidoptera. *Heredity* 9:323–42.

56. E.B.F. to H.B.D.K. 1 July 1953. H.B.D. Kettlewell papers. Bodleian Library. 'I am glad to hear that you have so satisfactory a place in which to work. It is disappointing that the recoveries are not better and curious that you are getting 10% of the typical betularia. However, I do not doubt that the results will be very well worth while ... I am sorry to hear of your lumbago and hope it quickly clears up.'

57. Steve Jebson of the UK Meteorological Office supplied Birmingham (Elmdon) Airport weather data for 25 June–5 July 1953.

58. H.B.D. Kettlewell. 1955. Selection experiments on industrial melanism in the Lepidoptera. Op. cit.

59. Ibid.

60. E.B.F. to H.B.D.K. 24 September 1953. H.B.D. Kettlewell papers. Bodleian Library.

61. E.B.F. to H.B.D.K. 17 November 1953. H.B.D. Kettlewell papers. Bodleian Library. 'When I wrote the chapter on Melanism last February, you had not, of course, cleared up one of the mysteries of that subject ... Now at last we have got definite evidence of the activity of birds in eliminating these insects, and doing so differentially, and it is clear why a phenomenon, which must in reality be so common and widespread, has heretofore escaped notice. Will you allow me to scrap the passage ... and give a brief account of your own results in its place?'

62. E.B. Ford. 1955. *Moths*. No. 30. New Naturalist Series. London: Collins.

63. H.B.D.K. to W. Bowater. 24 September 1954. H.B.D. Kettlewell papers. Bodleian Library.

64. H.B.D. Kettlewell. 1955. The life history of *Hydracea hucherardi mabile* (Lep., Agrotidae). *The Entomologist* 88 (1109): 218–19. October 1955.

65. P.B.M. Allan. 1956. A review of E.B. Ford's *Moths. Entomologist's Record and Journal of Variation* 67: 104.

66. H.B.D.K. to P.B.M. Allan. 16 March 1955. H.B.D. Kettlewell papers. Bodleian Library.

67. P.B.M. Allan to H.B.D.K. 1 April 1955.

68. On 8 March 1955 H.B.D.K. wrote to E.A. Cockayne: 'You must have realized . . . what a big influence you were in my life, and how much I admired you both as a man and in the work you did. It is a great disillusionment . . . that you have turned out as you have.' H.B.D. Kettlewell papers. Bodleian Library.

69. H.B.D.K. to Cyril A. Clarke. 18 March 1955. H.B.D. Kettlewell papers. Bodleian Library.

70. R.J. (Sam) Berry is now professor of biology at University College, London.

71. Interview with Kate Davies.

72. H.B.D.K. to C.A.C. 26 May 1955. H.B.D. Kettlewell papers. Bodleian Library.

73. Most of the details can be found in a letter from H.B.D.K. to Leslie Goodson, 28 July 1955. H.B.D. Kettlewell papers. Bodleian Library.

74. Ibid.

75. E.B.F. to H.B.D.K. 11 July 1955. The letter reads, in part: 'I am not at all satisfied by this [H.B.D.K.'s medical situation] until you have seen a specialist here, or in London . . . it is perfectly clear that you have been overdoing things . . . moving into your new house. I absolutely insist that you take things quietly over the next few weeks.'

76. Niko Tinbergen. 1979. Happy moments with Bernard Kettlewell. *Lycidas*. Wolfson College, Oxford.

77. H.B.D. Kettlewell. Further selection experiments on industrial melanism in the Lepidoptera. *Heredity* 10(3): 287–301. December 1956.

78. Niko Tinbergen to H.B.D.K. 15 July 1955.

79. H.B.D.K. to Leslie Goodson. 28 July 1955. H.B.D. Kettlewell papers. Bodleian Library. '. . . if this doesn't satisfy P.B.M. Allan

nothing will . . . I think he richly deserves it, and, as I am already booked for showing it in five places, it should eventually get round to him!'

80. E.A.C. to H.B.D.K. 23 September 1955. H.B.D. Kettlewell papers. Bodleian Library.

81. Stephen Jay Gould. 1983. The hardening of the Modern Synthesis. In Marjorie Grene (ed.). *Dimensions of Darwinism: Themes and Counterthemes in Twentieth-Century Evolutionary Theory*. Cambridge: Cambridge University Press.

82. Julian S. Huxley. 1953. *Evolution in Action*. New York: Harper.

83. Edward O. Wilson *et al.* 1973. *Life on Earth*. Stamford, Conn.: Sinauer Associates.

84. Interview with William B. Provine.

85. Interview with Audrey Z. Smith.

86. H.B.D. Kettlewell. 1958. A survey of the frequencies of *Biston betularia* L. (Lep.) and its melanic forms in Britain. *Heredity* 12: 51–72.

87. H.B.D.K. to F.W.J. 1955. H.B.D. Kettlewell papers. Bodleian Library.

88. Interview with Kate Davies.

89. I am indebted to Ruth Wickett for showing me this ditty, authored by 'H.U. Marr'.

90. Professor J. James Murray, now Samuel Miller Professor of Biology at the University of Virginia, attended Oxford as a Rhodes scholar in the early 1950s, and later as a doctoral student, earning his doctorate in 1962.

91. J.W. Heslop Harrison. 1956. Melanism in the Lepidoptera. *Entomologist's Record and Journal of Variation* 68:172–81. 15/VIII/56.

92. H.B.D. Kettlewell. 1956. Melanism and an answer to J.W. Heslop Harrison. *Entomologist's Rec. and J. Var.* 68. 15/XII/56.

93. Kurt Sabbagh. *A Rum Affair*. Op. cit.

94. E.B.F. to H.B.D.K. 31 August 1956. H.B.D. Kettlewell papers. Bodleian Library.

95. Vassiliki Betty Smocovitis. 1996. *Unifying Biology: The Evolutionary*

Synthesis and Evolutionary Biology. Princeton, NJ: Princeton University Press.

96. Stephen Jay Gould. 1996. The bare bones of natural selection. In *Full House: the Spread of Excellence from Plato to Darwin*. New York: Harmony Books.

97. Interview with Austin P. Platt, emeritus professor of biological sciences at the University of Maryland Baltimore County (UMBC).

98. Stephen Jay Gould. 1983. The hardening of the Modern Synthesis. Op. cit.

99. R.H. MacArthur and J.H. Connell. 1966. *The Biology of Populations*. New York & London: John Wiley and Sons.

100. A thorough discussion of the centennial appears in V.B. Smocovitis. 1999. The 1959 Darwin Centennial Celebration in America. *Osiris* 14:274–323.

101. Sol Tax and Charles Callender (eds). 1960. *Evolution After Darwin: The University of Chicago Centennial*. Chicago: University of Chicago Press.

102. V.B. Smocovitis. The 1959 Darwin Centennial Celebration in America. Op. cit.

103. H.B.D.K. to P.M.S. 14 January 1964. Philip Sheppard papers. American Philosophical Society, Philadelphia.

104. P.M.S. to H.B.D.K. 10 January 1964. Philip Sheppard papers. A.P.S.

105. H.B.D.K. to J.B.S. Haldane. 3 June 1963. H.B.D. Kettlewell papers. Bodleian Library.

106. H.B.D. Kettlewell. 1965. A 12-year survey of the frequencies of *Biston betularia* L. and its melanic forms in Britain. *Entomologist's Rec. & J. Var.* 77: 195–218.

107. P.M.S. to H.B.D.K. 5 November 1965; H.B.D.K. to E.B.F. 9 November 1965. H.B.D. Kettlewell papers. Bodleian Library.

108. E.B.F. to P.M.S. 17 April 1963. Philip Sheppard papers. A.P.S.

109. Interview with David Kettlewell.

110. E.B.F. to P.M.S. 30 May and 1 June 1961. Philip Sheppard papers. A.P.S.

111. J.B.S. Haldane. 1964. MS. E.B. Ford papers. Bodleian Library.

112. J.B.S.H. to H.B.D.K. 29 May 1956. H.B.D. Kettlewell papers. Bodleian Library.

113. Interview with J.J. Murray; and Bryan C. Clarke, 'Edmund Brisco Ford', *Biographical Memoirs of Fellows of the Royal Society*. Vol. 41 (1995), p. 52.

114. R.J. Berry. 1990. Appendix C. Industrial melanism and peppered moths (*Biston betularia* (L.)). *Biological Journal of the Linnean Society* 39:319–22.

115. H.B.D.K. to Cyril A. Clarke. November 1963. H.B.D. Kettlewell papers. Bodleian Library.

116. H.B.D. Kettlewell. 1965. Insect survival and selection for pattern. *Science* (New York) 148: 1290–5.

117. R.J. Berry. 1990. Industrial melanism and peppered moths (*Biston betularia* (L.)). *Biol. J. Linn. Soc.* 39:301–22.

118. The papers of H.B.D. Kettlewell were donated to Wolfson College after his death.

119. An account of the meeting is found in P.S. Moorhead & M.W. Kaplan. 1967. *Mathematical Challenges to the Neo-Darwinian Interpretation of Evolution*. Philadelphia: Wistar Institute Press.

In another context Waddington asserted: 'Natural selection is that some things leave more offspring than others; and you ask, which leave more offspring than others? And it is those that leave more offspring; and there is nothing more to it than that. The whole guts of evolution – which is, how do you come to have horses and tigers and things – is outside the mathematical theory.' (Tom Bethell. Darwin's Mistake. *Harper's*. February 1976.)

120. H.B.D.K. to Julian Huxley. 6 February 1964. H.B.D. Kettlewell papers. Bodleian Library.

121. E.B.F. to H.B.D.K. March 1966. H.B.D. Kettlewell papers. Bodleian Library.

122. G. Hardin. 1966. *Biology: Its Principles and Implications*. San Francisco: W.H. Freeman and Company.

123. James Cadbury, an ornithologist, was the son of the Cadburys who permitted Bernard to carry out his experiments on their property near Birmingham.

124. I am indebted to R.J. Berry and the late David Kettlewell for their descriptions of the expeditions to the Shetland Islands. For more information on the research see H.B.D. Kettlewell and R.J. Berry. 1969. Gene flow in a cline: *Amathes glareosa* Esp. and its melanic f. *edda* Stdg. (Lep.) in Shetland. *Heredity* (London) 14: 1–14.

125. P.M.S. to E.B.F. 18 May 1970. Philip Sheppard papers. American Philosophical Society, Philadelphia.

126. P.M.S. to H.B.D.K. 6 October 1966. H.B.D. Kettlewell papers. Bodleian Library. 'I think that how one interprets the reduction in the frequency of carbonaria between 1962 and 1965 is very much a matter of opinion. Incidentally, the frequency is still down this year. The point I was trying to bring home was that the change in frequency appeared far too rapid for a change in gene frequency. It virtually occurred in one year, or at the most two, which would mean changes in selective value of a very high magnitude. Moreover, it would mean that subsequently there had been some other change which kept the population at the lower frequency and did not cause it to continue to decline or to rise again.'

127. C.A. Clarke and P.M. Sheppard. 1966. A local survey of the distribution of industrial melanic forms in the moth *Biston betularia* and estimates of the selective values of these in an industrial environment. *Proceedings of the Royal Society of London B* 263: 35–70.

128. Letters between E.B.F. and P.M.S. April 1968. Philip Sheppard papers. A.P.S.

129. Interview with Jim Kettlewell.

130. H.B.D.K. to A.B. Klots. 2 July 1968. H.B.D. Kettlewell papers. Bodleian Library.

131. Larry Gilbert of the University of Texas.

132. J.B.S. Haldane. 1957. The cost of natural selection. *Journal of Genetics* 55:511–24.

133. Creationists are fond of quoting from Walter ReMine's *The Biotic Message*, which argues that human beings and chimps differ in 2 to 3 per cent of the genome, but that to alter just 0.014 per cent of the genome by means of point mutations (nucleotide substitutions) would require 500,000 generations, or ten million years – two to three times longer than the date of the divergence of human and ape ancestral lines.

134. R.C. Lewontin and J.L. Hubby. 1966. A molecular approach to the study of genic heterozygosity in natural populations. II. Amount of variation and degree of heterozygosity in natural populations of *Drosophila pseudoobscura*. *Genetics* 54:595–609.

135. In his 1983 book *The Neutral Theory of Evolution* (New York: Cambridge University Press), Motoo Kimura argued that 'the picture of evolutionary change that actually emerged' was 'quite incompatible with the expectations of neo-Darwinism'. Even using conservative estimates, the overall rate of amino acid substitutions – that is, mutations – per genome per generation was very high. 'I then realized,' he wrote, 'that this estimate . . . is at least several times higher for mammals than the famous estimate previously obtained by Haldane (1957).' What Haldane had called the 'cost of natural selection' became Kimura's 'substitutional load', and Kimura calculated that if the mutations were under the control of natural selection, 'the substitutional load in each generation is so large that no mammalian species could tolerate it.'

The majority of nucleotide substitutions in the course of evolution must be the result not of Darwinian selection but of random fixation of selectively neutral or nearly neutral mutants, Kimura insisted in 1968 (M. Kimura. 1968. Evolutionary rate at the molecular level. *Nature* 217: 624–6.) Lester King and Thomas H. Jukes independently arrived at the same idea and wrote an influential paper in 1969. (J.L. King and T.H. Jukes. 1969. Non-Darwinian evolution: random fixation of selectively neutral mutations. *Science* 164: 788–98.)

136. Letter from E.B. Ford to Th. Dobzhansky. 26 May 1969. E.B. Ford papers. Bodleian Library.

137. E.B. Ford papers. Bodleian Library. In a June 1968 letter to Kennedy McWhirter he complained:

> It is extraordinary that this business of neutral genes has raised its head again. It seems to me that the mathematics of Fisher's brilliant paper in the Proceeding of the Royal

Society of Edinburgh, in 1930, surely still hold. You remember they show how accurate must be the balance of advantage and disadvantage between the two alleles if they are to be selectively neutral relative to one another; consequently such a situation must be rare . . .

Bearing in mind the well known calculation of the rate of spread of a neutral gene it is evident that the two alleles could not have reached the frequencies that are being found if they were spreading from mutation as rare as we know mutation to be. If mutation were much commoner, it would threaten the stability of the genetic material . . . It is to be noticed that people like Kimura entirely fail to take into account the rate of evolution as ascertained in nature and the force of selection . . .

138. J.R.G. Turner. 1992. Stochastic processes in populations. Op. cit.

139. 'Edmund Brisco Ford'. Memorial address by Bryan Wilson, delivered in the chapel of All Souls College. 12 March 1988.

140. E.B.F. to Professor E. Bosiger. 25 July 1969. E.B. Ford papers. Bodleian Library.

141. E.B.F. to H.B.D.K. 23 July 1969. H.B.D. Kettlewell papers. Bodleian Library

142. E.B.F. to P.M.S. 7 July 1969. Philip Sheppard papers. A.P.S.

143. H.B.D.K. to J.S.H. 24 May 1973. H.B.D. Kettlewell papers. Bodleian Library.

144. H.B.D.K. to P.M.S. June 1972. Philip Sheppard papers. A.P.S. Despite his incapacitation, Bernard still had moths on the brain, writing to Sheppard: 'I urgently need Biston betularia back-cross heterozygous pairings – any possibility of your being able to provide eggs of such a mating?'

145. E.B.F. to H.B.D.K. August 1973. Right after this 'review', Ford added: 'It was most kind and helpful of you to discuss my "issue of blood."' H.B.D. Kettlewell papers. Bodleian Library.

146. P.M. Sheppard. 1973. Pigmentation and Evolution. *Nature* 246: 535–536 December 21/28 1973.

147. P.M.S. to H.B.D.K. 16 October 1973. Philip Sheppard papers. A.P.S.

148. H.B.D.K. to P.M.S. 26 October 1973. Philip Sheppard papers. A.P.S.

149. H.B.D.K. to P.M.S. 23 January 1974. Philip Sheppard papers. A.P.S.

150. E.B.F. to H.B.D.K. 31 August 1956. H.B.D. Kettlewell papers. Bodleian Library.

151. Interview with Bryan Clarke.

152. E.B.F. to J.J. Murray. 20 January 1972. J.J. Murray.

153. Richard Lewontin. 1972. Testing the theory of natural selection. *Nature* 236: 24 March 1972.

154. MS. E.B. Ford papers. Bodleian Library.

155. R.C. Lewontin. 1974. *The Genetic Basis of Evolutionary Change.* New York: Columbia University Press.

156. J.A. Bishop. 1972. An experimental study of the cline of industrial melanism in *Biston betularia* (L) (Lepidoptera) between urban Liverpool and rural north Wales. *Journal of Animal Ecology* 41: 209–43.

157. D.R. Lees and E.R. Creed. 1975. Industrial melanism in *Biston betularia*: the role of selective predation. *Journal of Animal Ecology* 44: 67–83.

158. J.A. Bishop and L.M. Cook. 1975. Moths, melanism, and clean air. *Scientific American* 232: 90–9.

159. Interview with David Kettlewell.

160. E.B.F. to P.M.S. 2 August 1976. Philip Sheppard papers. A.P.S.

161. Interview with Miriam Rothschild.

162. H.B.D.K. to J.S.H. 11 May 1965. H.B.D. Kettlewell papers. Bodleian Library.

163. Various letters from H.B.D.K. to Cyril A. Clarke and Miriam Rothschild, 1977. H.B.D. Kettlewell papers.

164. H.B.D.K. to C.A.C. 14 February 1979. H.B.D. Kettlewell papers. Boldleian Library. On 19 March, Clarke wrote to H.B.D.K.: 'I spoke to Charles [his son, a doctor] last night and you are definitely on the waiting list . . . a delay may be as long as 2 months.'

165. J.R.G. Turner. 1980. 'Kettlewell, Henry Bernard Davis', pp. 459–71 in F. Holmes (ed.). *The Dictionary of Scientific Biography.* New York: Charles Scribners Sons.

166. Interview with John R.G. Turner.

167. Cyril Clarke. 1979. Obituary: Dr Henry Bernard Davis Kettlewell. *Entomologist's Rec. and J. Var.* I/X/79.

168. Interview with John Haywood.

169. S.J. Gould and R.C. Lewontin. 1979. The spandrels of San Marco and the Panglossian paradigm: a critique of the adaptationist program. *Proc. Roy. Soc. Lond. B* 205: 581–98.

170. On another occasion, at a 1987 meeting in Basel, Cain gave a plenary address in which he vehemently defended his land snails and denounced Gould and Lewontin. 'Had we adopted the exhortations of Gould and Lewontin (1979) to abandon what they caricature as the adaptationist programme, it would have been the finest possible recipe for doing nothing, and finding out nothing. (A.J. Cain. 1988. Evolution. *Journal of Evolutionary Biology* 1: 185–94.)

171. E.B. Ford. 1980. *Taking Genetics into the Countryside*. London: Weidenfeld and Nicolson.

172. E.B. Ford. The Abilities of the Sexes, and the Admission of Women to All Souls (Confidential to the Warden and Fellows of All Souls). MS. E.B. Ford papers. Bodleian Library.

173. c.f. Bryan C. Clarke. Edmund Brisco Ford. Op. cit.

174. Ibid.

175. E.B.F. to P.M.S. 14 June 1968. Philip Sheppard papers. A.P.S. 'Naturally we all want his views to be perpetuated, but I am not really happy about these volumes of collected works. Even with Fisher and so much with other people, what a man says when he is young may not be what he would have wished in later years. Nor am I quite happy about biography. It is in his work and thoughts, rather than in himself, that the true interest lies. However, the daughter has picked upon certain people . . . to record memories of him. This I did to a magnetic tape sent to me for that purpose (I hope to you also). So I sat in front of it and just talked: more or less without preparation. I hope I may have said something to bring his strange personality to mind, but I doubt it for that is not easy.'

176. MS. Questionnaire Concerning the Evolutionary Synthesis. E.B. Ford papers. Bodleian Library.

177. Interview with J. James Murray.

178. Bryan C. Clarke. Edmund Brisco Ford. Op. cit.

179. Interview with Bryan Clarke.

180. H.B.D. Kettlewell. 1955. Recognition of appropriate backgrounds by the pale and black phases of Lepidoptera. *Nature* (London) 175: 943–4.

181. H.B.D. Kettlewell and D.L.T. Conn. 1977. Further background-choice experiments on cryptic Lepidoptera. *J. Zool.* (London) 181: 371–6.

182. T.D. Sargent. 1969. Background selections of the pale and melanic forms of the cryptic moth, *Phigalia titea* (Cramer). *Nature* (London) 222: 585–6.

Later, other scientists would also fail to replicate Kettlewell's results. In 1979 Kauri Mikkola of the University of Helsinki found no difference between the different phenotypes in their choice of blackish or whitish backgrounds. In 1988 Rory Howlett of Cambridge and Bruce Grant of Virginia's William and Mary College reported that peppered moths' preferences for background appeared unrelated to their phenotype.

183. Douglas J. Futuyma. 1979. *Evolutionary Biology*. Sunderland, Mass.: Sinauer Associates.

184. More details can be found in the following papers:

T.R. Manley. 1981. Frequencies of the melanic morph of *Biston cognataria* (Geometridae) in a low-pollution area of Pennsylvania from 1971 to 1978. *Journal of the Lepidopterists' Society* 35: 257–65.

D.F. Owen. 1962. Industrial melanism in North American moths. *American Naturalist* 95: 227–33.

T.D. Sargent. 1983. Melanism in *Phigalia titea* (Cramer) (Lepidoptera: Geometridae): a fourteen-year record from central Massachusetts. *Journal of the New York Entomological Society* 91: 75–82.

D.A. West. 1977. Melanism in *Biston* (Lepidoptera: Geometridae) in the rural central Appalachians. *Heredity* 39: 75–81.

185. T.D. Sargent. 1976. *Legion of Night: The Underwing Moths*. Op. cit.

186. R.H. Brady. 1982. Dogma and doubt. *Biol. J. Linn. Soc.* 17: 79–96.

187. K. Mikkola. On the selective forces acting in the industrial melanism of *Biston* and *Oligia* moths (Lepidoptera: Geometridae and Noctuidae). First published in 1979 in *Annales Entomologicai Fennici* 45: 81–7; republished in 1984 in *Biological Journal of the Linnean Society* 21: 409–21.

188. J.P. Hailman. 1982. Evolution and Behaviour: an Iconoclastic View. In H.C. Plotkin (ed.). *Learning, Development, and Culture*, pp. 205–54. London: John Wiley and Sons.

189. J.S. Jones. 1982. More to melanism than meets the eye. *Nature* (London) 300: 109–10.

190. See G.S. Mani. 1982. A theoretical analysis of the morph frequency variation in the peppered moth over England and Wales. *Biol. J. Linn. Soc.* 17: 259–67. And G.S. Mani. 1990. Theoretical models of melanism in *Biston betularia* – a review. *Biol. J. Linn. Soc.* 39: 355–71.

191. T.G. Liebert and P.M. Brakefield. 1987. Behavioural studies of the peppered moth *Biston betularia* and a discussion of the role of pollution and lichens in industrial melanism. *Biol. J. Linn. Soc.* 31: 129–50.

192. P.D.J. Whittle *et. al.* followed the same experimental procedures as Clarke and Sheppard (1966), placing frozen specimens of both types on appropriate or contrasting backgrounds, or placing them randomly. At an unpolluted site neither method yielded evidence of selective predation, while at the polluted site *carbonaria* apparently enjoyed a cryptic advantage. (P.D.J. Whittle, C. Clarke, P.M. Sheppard and J.A. Bishop. 1976. Further studies on the industrial melanic moth *Biston betularia* (L.) in the northwest of the British Isles. *Proc. Roy. Soc. Lond. B* 194: 467–80.)

In 1977 R.C. Steward gauged the 'relative crypsis' of the phenotypes of *Biston betularia* at 52 sites in southern England and compared these values to their frequencies. He found that crypsis could account for only 18 per cent of the variation in *carbonaria* frequencies at different sites. In southern Britain, he concluded, selective predation is of secondary importance, and some other factor chiefly determines the frequencies of the phenotypes of the peppered moth. (R.C. Steward. 1977. Melanism and selective predation in three species of moths. *Journal of Animal Ecology* 46: 483–96.)

In 1980, Murray *et. al.* repeated Bishop's procedure of 1972 and found no significant differences in the predation on the morphs of *Biston betularia*. (N.D. Murray, J.A. Bishop and M.R. Macnair. 1980. Melanism and predation by birds in the moths *Biston betularia* and *Phigalia pilosaria*. *Proc. Roy. Soc. Lond. B:* 210: 277–83.)

In a predation experiment at two different sites, in 1987, R.J. Howlett and M.E.N. Majerus glued dead *typica* and *carbonaria* either fully exposed on tree trunks or placed just under the joint between the trunk and a major branch. Their results supported Kettlewell's report of differences in survival values for the two morphs (more *typica* eaten in the polluted region, more *carbonaria* in the unpolluted regions). However, the predation was significantly greater for moths exposed on the trunk than for those below the trunk/branch joint – a finding corroborated in 1993 by Carys Jones, who found that moths under lateral branches were about half as likely to be eaten. (R.J. Howlett and M.E.N. Majerus. 1987. The understanding of industrial melanism in the peppered moth (*Biston betularia*) (Lepidoptera: Geometridae). *Biol. J. Linn. Soc.* 30: 31–44.)

193. J.A. Bishop and L.M. Cook. 1975. Moths, melanism, and clean air. Op. cit.

194. E.R. Creed, D.R. Lees and M.G. Bulmer. 1980. Pre-adult viability differences of melanic *Biston betularia* (L) (Lepidoptera). *Biol. J. Linn. Soc.* 30: 31–44.

195. Craig Holdrege. The case of the peppered moth illusion. Op. cit.

196. R.J. Berry. 1990. Industrial melanism and peppered moths (*Biston betularia* (L.)). *Biol. J. Linn. Soc.* 39: 301–22.

The list of 'legitimate criticisms', 'gaps in knowledge' and so on can be found in Appendix B, by Berry, in the same volume.

Also in the same volume, the article by G.S. Mani, 'Theoretical models of melanism in *Biston betularia* – a review', lists major problems and inconsistencies in contemporary theoretical models.

197. M.E.N. Majerus. *Melanism: Evolution in Action.* Op. cit.

198. Bruce Grant. 1999. Fine tuning the peppered moth paradigm. *Evolution* 53: 980–4. The passage continues: 'Such moths could then select their own hiding places and would either survive or fail to survive the following day before trapping begins. This solution might seem obvious and easy to recommend, but it's not easy to do. The return rate two days after release drops off enormously as a result of either mortality or dispersal from the trapping area. To get enough data for statistical comparisons requires releasing many moths, and to rear them from the egg stage requires feeding fresh leaves to thousands of caterpillars the year preceding the experiment. Few who haven't fed large numbers of growing caterpillars can appreciate how labor-intensive the task is. Still, the experiment should be done.'

199. Interview with Jerry Coyne (Dick Teresi).

200. M.E.N. Majerus. *Melanism: Evolution in Action.* Op. cit.

201. Ibid.

202. D.R. Lees and E.R. Creed. Industrial melanism in *Biston betularia*: the role of selective predation. Op. cit.

203. D.R. Lees. 1981. Industrial melanism: genetic adaptation of animals to air pollution. In Laurence Cook, *et al. Genetic Consequences of Man Made Change*, pp. 129–76. New York: Academic Press.

204. E.R. Creed, D.R. Lees and M.G. Bulmer. Pre-adult viability differences of melanic *Biston betularia* (L) (Lepidoptera). Op. cit.

205. He wrote to a lepidopterist friend in 1959: 'No one would be foolish enough to argue that your statement "The greatest enemies of moths are not the birds but the bats" is untrue, but bats hunt auditorially ... and therefore their predation is N O T S E L E C-T I V E. It does not matter the slightest if bats take 90% of a species population at random on the wing but if birds which hunt visually account for the other 10%, but do this S E L E C T I V E L Y, this and this alone will be reflected in the speed at which the more disadvantageous form spreads through the population.' H.B.D.K. to B.J. Lempke. 24 June 1959. H.B.D. Kettlewell papers. Bodleian Library.

206. C.A. Clarke, B. Grant, F.M.M. Clarke and T. Asami. 1994. A long term assessment of *Biston betularia* (L.) in one U.K.

locality (Caldy Common near West Kirby, Wirral), 1959–1993, and glimpses elsewhere. *Linnean* 10 (2): 18–26.

207. B.S. Grant, D.F. Owen and C.A. Clarke. September/October 1996. Parallel rise and fall of melanic peppered moths in America and Britain. *Journal of Heredity* 97: 351–7.

208. Carol Kaesuk Yoon. 1996. Parallel Plots in Classic of Evolution. *The New York Times*: p. C-1. 12 November 1996.

209. Michael E.N. Majerus. *Melanism: Evolution in Action.* Op. cit.

210. L.M. Cook, K.D. Rigby and M.R.D. Seward. 1990. Melanic moths and changes in epiphytic vegetation in north-west England and north Wales. *Biol. J. Linn. Soc.* 39: 343–54.

211. Interview with Bruce Grant; and B.S. Grant and L.L. Wiseman, 'Recent history of melanism in American peppered moths', in press. In this paper the authors make the following intriguing observation: 'We would caution here that our present study, taken alone, might be interpreted as evidence for thermal melanism, a phenomenon not uncommon in the Lepidoptera (Majerus 1998). However, there is direct evidence against thermal melanism in *Biston betularia* [in Britain and Scandinavia] because of the absence of latitudinal clines.'

212. An account of this research is found in T.D. Sargent. Industrial melanism in moths: a reassessment. In M. Wicksten (compiler). *Adaptive Coloration in Invertebrates.* College Station, Texas: Texas A & M University.

213. Theodore D. Sargent, Craig D. Millar and David Lambert. 1998. The 'classical' explanation of industrial melanism. In *Evolutionary Biology*, vol. 30, Max K. Hecht *et al.* (eds). New York: Plenum Press.

214. Interview with Jerry Coyne (Dick Teresi).

215. Jerry A. Coyne. 1998. Not black and white. *Nature* 396: 35–6. 5 November 1998.

216. Jonathan Wells. 1999. Second thoughts about peppered moths. *The Scientist.* 24 May 1999.

217. Robert Matthews. 1999. Scientists pick holes in Darwin's moth theory. *The Daily Telegraph.* London. 18 March 1999.

218. Interview with Douglas Futuyma (Dick Teresi).

219. c.f. H.B.D. Kettlewell. 1943–4. Temperature experiments on the pupae of *Heliothis peltigera* Schiff. and *Panaxia dominula* L. *Proc. S. Lond. ent. nat. hist. soc.*: 69–81.

220. Endler explained: 'If the tendency to induce is inherited as a single gene – if the moth eats the host plant and immediately turns black – it could work as a straight Mendelian character. It could be a threshold character; the melanic form would have a lower threshold of induction than the typical.'

221. c.f. studies by S.W. Bromley (southern New England), C.G. Lorimer (northeastern Maine), T.G. Siccama (northern Vermont).

222. See T.D. Sargent. Industrial melanism in moths. Op. cit.

223. Michael Balter. Was Lamarck just a little bit right? 7 April 2000. *Science* 288: 38.

224. In 1957 Conrad Waddington described a process called 'genetic assimilation' wherein an environmentally induced change in an organism may occasionally become genetically fixed. The most celebrated example was the evolution of mutant fruit flies without crossveins in response to heat shock. More recently, experiments with tobacco plants and flax have demonstrated genetic change through the effects of fertilizers.

225. L.M. Cook. 2000. Changing views on melanic moths. *Biol. J. Linn. Soc.* 69: 31–41.

226. Richard C. Lewontin. 1972. Testing the theory of natural selection. *Nature* 236: 181–2.

In their famous 'Spandrels' paper, Stephen Jay Gould and Lewontin lodged the same complaint against the 'adaptationist programme', affirming that they would object less strenuously to it 'if its invocation, in any particular case, could lead in principle to its rejection for lack of evidence'.

227. L.M. Cook. Changing views on melanic moths. Op. cit.

228. For example, working with guppies from streams in Trinidad and Venezuela, Endler employed careful standardized techniques to score the colouring of their spots, to divide streams into sections, to keep track of the predators, to take periodic censuses of the survivors, and so on. In some experiments he used tanks with different mixes of gravel and added or subtracted predators, comparing the evolution of the fish swimming in tanks without predators to those in tanks with predators. In other tests he was able to analyse in a highly systematic way the opposing forces of sexual selection (bright spots favoured successful mating) and survival (dimmer spots worked better as camouflage, protecting the fish from predators).

229. Adrian W. Wenner and Patrick H. Wells. 1990. *Anatomy of a Controversy: The Question of a 'Language' Among Bees*. New York: Columbia University Press.

230. Karl von Frisch's theory has since been vindicated, but Wenner's point, that the data could equally fit another hypothesis, was worthy of consideration.

231. Craig Holdrege. The case of the peppered moth illusion. Op. cit.

232. It is worth noting that industrial melanism remained an open-ended investigation for Bernard Kettlewell himself. While textbook accounts treat the 1953–5 experiments as the 'end of the story', according to David Rudge, a philosopher of science at Western Michigan University, to Kettlewell they represented a beginning. He proceeded to carry out extensive surveys and to 'research every aspect of the phenomenon of melanism', from background choice to the evolution of dominance. In Rudge's view, in the preface to his book 'Kettlewell describes the 423-page account as "incomplete" . . . Clearly [he] did not see the mark–release–recapture experiments as the end of the story.' David Wyss Rudge. 1999. Taking the peppered moth with a grain of salt. *Biology and Philosophy* 14: 9–37.

233. Arthur M. Shapiro. 2000. Review of *Melanism: Evolution in Action* by Michael E.N. Majerus. *Journal of the Lepidopterists' Society* 45 (1): 38.

234. Isaac Asimov. 1984. *Asimov's New Guide to Science*. New York: Basic Books.

235. Arthur M. Shapiro. Review of *Melanism: Evolution in Action*. Op. cit.

236. Michael Ruse. 1999. *Mystery of Mysteries: Is Evolution a Social Construction?* Cambridge, Mass.: Harvard University Press.

237. H.B.D.K. to P. Harper. 16 October 1970. H.B.D. Kettlewell papers. Bodleian Library.

238. H.B.D.K. to A.B. Klots. December 1974. H.B.D. Kettlewell papers. Bodleian Library.

Bibliography

Books

Allan, P.B.M. 1957. *A Moth-Hunter's Gossip*. London: Watkins & Doncaster.

Allan, Garland E. 1975. *Life Science in the Twentieth Century*. New York: John Wiley and Sons.

Baer, Adela S. *et al.* 1971. *Central Concepts of Biology*. New York: Macmillan.

Behe, Michael J. 1996. *Darwin's Black Box: The Biochemical Challenge to Evolution*. New York: Touchstone/Simon & Schuster.

Berry, R.J., T.J. Crawford and G.M. Hewitt (eds). 1992. *Genes in Ecology*: The 33rd Symposium of the British Ecological Society, University of East Anglia, 1991. Oxford: Blackwell Scientific Publications.

Bishop, J.A. and L.M. Cook (eds). 1981. *Genetic Consequences of Man Made Change*. London: Academic Press.

Bowler, Peter J. 1984. *Evolution: The History of an Idea*. Berkeley: University of California Press.

Box, Joan Fisher. 1978. *R.A. Fisher: The Life of a Scientist*. New York: John Wiley and Sons.

Clark, Ronald William. 1984. *J.B.S.: The Life and Work of*

J.B.S. Haldane. Oxford: Oxford University Press.

Creed, Robert (ed.). 1971. *Ecological Genetics and Evolution: Essays in Honour of E.B. Ford*. Oxford: Blackwell.

Darwin, Charles. 1958 (1859). *On the Origin of Species*. New York: New American Library.

Darwin, Francis (ed.). 1911. *The Life and Letters of Charles Darwin, Including an Autobiographical Chapter*. New York, London: D. Appleton & Co. Vol. 3.

Dawkins, Richard. 1986. *The Blind Watchmaker*. London: Longman.

Dobzhansky, Theodosius. 1937. *Genetics and the Origin of Species*, 1st edition. New York: Columbia University Press.

———, Francisco J. Ayala, Ledyard G. Stebbins and James W. Valentine. 1977. *Evolution*. San Francisco: W.H. Freeman & Company.

Dunn, L.C. 1956. *A Short History of Genetics*. New York: McGraw-Hill.

Eldredge, Niles. 1999. *The Pattern of Evolution*. New York: W.H. Freeman and Company.

———. 2000. *The Triumph of Evolution*. New York: W.H. Freeman and Company.

Endler, John A. 1986. *Natural Selection in the Wild*. Princeton, NJ: Princeton University Press.

Fabre, Jean Henri. 1916 (1915). *The Life of a Caterpillar*. Tr. by Alexander Teixera de Mattos. New York: Dodd Mead and Co.

Fisher, R.A. 1930. *The Genetical Theory of Natural Selection*. Oxford: Clarendon Press.

Ford, Edmund Brisco. Papers. Bodleian Library. University of Oxford.

Ford, E.B. 1975. *Ecological Genetics* (4th edition). London: Chapman and Hall.

———. 1965. *Genetic Polymorphism*. London and Cambridge,

Mass.: Faber and Faber Ltd; MIT Press.

———. 1931. *Mendelism and Evolution*. London: Methuen and Co.

———. 1955. *Moths*. New Naturalist Series No. 30. London: Collins.

———. 1981. *Taking Genetics into the Countryside*. London: Weidenfeld and Nicolson.

Friedrich, Ekkehard. 1986. *Breeding Butterflies and Moths: A Practical Handbook for British and European Species*. Colchester, Essex, UK: Harley Books.

Futuyma, Douglas J. 1979. *Evolutionary Biology*. Sunderland, Mass.: Sinauer Associates Inc.

Gould, Stephen Jay. 1983. *Hen's Teeth and Horse's Toes*. New York: W.W. Norton & Company.

———. 1996. *Full House: The Spread of Excellence from Plato to Darwin*. New York: Harmony Books.

Grant, Verne. 1985. *The Evolutionary Progress: A Critical Review of Evolutionary Theory*. New York: Columbia University Press.

Grene, Marjorie (ed.). 1983. *Dimensions of Darwinism: Themes and Counterthemes in Twentieth-Century Evolutionary Theory*. Cambridge: Cambridge University Press.

Haldane, J.B.S. 1932. *The Causes of Evolution*. London: Longmans, Green and Co.

Hardin, G. 1966. *Biology: Its Principles and Implications*. San Francisco: W.H. Freeman and Company.

Himmelfarb, Gertrude. 1959. *Darwin and the Darwinian Revolution*. Garden City, NY: Doubleday.

Hitching, Francis. 1982. *The Neck of the Giraffe*. New York: Ticknor & Field.

Hofmann, H. and T. Marktanner. 1995. *Butterflies and Moths of Britain and Europe*. London: HarperCollins.

Holland, W.J. 1903. *The Moth Book*. New York: Doubleday.

Huxley, Julian. 1970. *Memories*. London: Allen & Unwin.

—— (ed.). 1940. *The New Systematics*. Oxford: Oxford University Press.

——. 1942. *Evolution: The Modern Synthesis*. London: Allen & Unwin.

Jastrow, Robert (general editor). 1984. *The Essential Darwin*. Boston: Little Brown.

Johnson, Kurt and Steve Coates. 1999. *Nabokov's Blues: The Scientific Odyssey of a Literary Genius*. Cambridge, Mass.: Zoland Books.

Jones, Steve. 1999, 2000. *Darwin's Ghost: The Origin of Species Updated*. New York: Random House.

Kettlewell, Henry Bernard Davis. Papers. Bodleian Library. University of Oxford. (Collection owned by Wolfson College, Oxford.)

Kettlewell, Bernard. 1973. *The Evolution of Melanism: The Study of a Recurring Necessity*. Oxford: Clarendon Press.

Kimura, Motoo. 1983. *The Neutral Theory of Evolution*. New York: Cambridge University Press.

Kingsland, Sharon E. 1985. *Modeling Nature: Episodes in the History of Population Ecology*. Chicago: University of Chicago Press.

Lewontin, Richard C. 1974. *The Genetic Basis of Evolutionary Change*. New York: Columbia University Press.

MacArthur, R.H. and J.H. Connell. 1966. *The Biology of Populations*. New York & London: John Wiley and Sons.

McNeill, J.R. 2000. *Something New Under the Sun*. New York: W.W. Norton & Company.

Majerus, Michael E.N. 1998. *Melanism: Evolution in Action*. London: Oxford University Press.

Mayr, Ernst. 1991. *One Long Argument: Charles Darwin and the Genesis of Modern Evolutionary Thought*. Cambridge, Mass.: Harvard University Press.

—— and William B. Provine. 1980, 1988. *The Evolutionary Synthesis: Perspectives in the Unification of Biology*. Cambridge, Mass.: Harvard University Press.

Moorhead, P.S. and M.W. Kaplan. 1967. *Mathematical Challenges to the Neo-Darwinian Interpretation of Evolution*. Philadelphia: Wistar Institute Press.

Morris, Jan. 1987, 1978. *Oxford*. New York: Oxford University Press.

Nabokov, Vladimir. 1958. *Lolita*. New York: Putnam's.

Paley, William. 1836. *Natural Theology*, vol. 1. London: Charles Knight.

Plotkin, H.C. (ed.). 1982. *Learning, Development, and Culture*. John Wiley & Sons.

Provine, William B. 1971. *The Origins of Theoretical Population Genetics*. Chicago: University of Chicago Press.

——. 1986. *Sewall Wright and Evolutionary Biology*. Chicago: University of Chicago Press.

Robson, G.C. and O.W. Richards. 1936. *The Variation of Animals in Nature*. London: Longmans Green.

Rose, Michael R. 1998. *Darwin's Spectre: Evolutionary Biology in the Modern World*. Princeton, NJ: Princeton University Press.

Ruse, Michael. 1998. *Darwinism Defended*. Reading, Mass.: Addison-Wesley.

——. 1999. *Mystery of Mysteries*. Cambridge, Mass.: Harvard University Press.

Sabbagh, Karl. 1999. *A Rum Affair*. London: Allen Lane/The Penguin Press.

Sargent, Theodore D. 1976. *Legion of Night: The Underwing Moths*. Amherst, Mass.: University of Massachusetts Press.

Sheppard, Philip Macdonald. Papers. American Philosophical Society. Philadelphia, Penn.

——. 1967. *Natural Selection and Heredity* (3rd ed.). London: Hutchinson University Library.

Smith, Audrey Z. 1986. *A History of the Hope Entomological Collections in the University Museum, Oxford.* Oxford: Clarendon Press.

Smith, John Maynard. 1966. *The Theory of Evolution* (2nd ed.). Baltimore, Maryland: Penguin Books.

Smocovitis, Vassiliki Betty. 1996. *Unifying Biology: The Evolutionary Synthesis and Evolutionary Biology.* Princeton, NJ: Princeton University Press.

Tax, Sol and Charles Callender (eds). 1960. *Evolution After Darwin: the University of Chicago Centennial*, vol. 3, *Issues in Evolution.* Chicago: University of Chicago Press.

Tutt, J.W. 1896. *British Moths.* London: George Routledge.

Ward, Jack A. and Howard R. Hetzel. 1980. *Biology: Today and Tomorrow.* St Paul, Minn.: West Publishing Company.

Weiner, Jonathan. 1995. *The Beak of the Finch.* New York: Vintage Books/Random House.

Wells, H.G., Julian Huxley and G.P. Wells. 1929. *The Science of Life.* London: Cassell & Co.

Wenner, Adrian H. and Patrick H. Wells. 1990. *Anatomy of a Controversy: The Question of a 'Language' Among Bees.* New York: Columbia University Press.

Williams, George C. 1966. *Adaptation and Natural Selection.* Princeton, NJ: Princeton University Press.

Wilson, Edward O. *et al.* 1973. *Life on Earth.* Stamford, CT: Sinauer Associates.

Articles

Allan, P.B.M. 1956. A review of EB Ford's *Moths. Entomologist's Record and Journal of Variation* 67: 104.

Allen, Garland E. 1979. Naturalists and experimentalists; the genotype and the phenotype. *Studies in the History of Biology* 3: 179–209.

Berry, R.J. 1990. Industrial melanism and peppered moths (*Biston betularia* (L.)). *Biological Journal of the Linnean Society* 39: 301–22.

—— and A.D. Bradshaw. Genes in the real world. In R.J. Berry, T.J. Crawford and G.M. Hewitt (eds), *Genes in Ecology*.

Bishop, J.A. 1972. An experimental study of the cline of industrial melanism in *Biston betularia* (L) (Lepidoptera) between urban Liverpool and rural north Wales. *Journal of Animal Ecology* 41: 209–43.

—— and L.M. Cook. 1975. Moths, melanism, and clean air. *Scientific American* 232: 90–9.

Brady, R.H. 1982. Dogma and doubt. *Biol. J. Linn. Soc.* 17: 79–96.

Brakefield, P.M. 1987. Industrial melanism: do we have the answers? *Trends in Ecology and Evolution* 2: 117–22.

——. 1990. A decline of melanism in the peppered moth *Biston betularia* in The Netherlands. *Biol. J. Linn. Soc.* 39: 327–34.

Burrian, Richard M. 1983. Adaptation. In Marjorie Grene (ed.), *Dimensions of Darwinism: Themes and Counterthemes in Twentieth-Century Evolutionary Theory*.

Cain, A.J. 1964. The perfection of animals. In J.D. Carthy and C.L. Duddington (eds), *Viewpoints in Biology 3*. London, Washington, DC: Butterworth's.

——. 1988. Criticism of J.R.G. Turner's article 'Fisher's evolutionary faith and the challenge of mimicry'. *Oxford Surveys of Evolutionary Biology* 5: 246–8.

—— and W.B. Provine. 1992. Genes and ecology in history. In R.J. Berry, T.J. Crawford and G.M. Hewitt (eds). *Genes in Ecology*, pp. 3–28.

———— and P.M. Sheppard. 1950. Selection in the polymorphic land snail *Cepaea nemoralis* (L.). *Heredity* 4: 275–94.

Cherfas, J. 1987. Exploding the myth of the melanic moth. *New Scientists* 25 December 1986/1 January 1987: 25.

Clarke, Bryan. 1995. Edmund Brisco Ford. *Biographical Memoirs of Fellows of the Royal Society* 41: 152.

Clarke, C.A. and P.M. Sheppard. 1966. A local survey of the distribution of industrial melanic forms in the moth *Biston betularia* and estimates of the selective values of these in an industrial environment. *Proceedings of the Royal Society of London B* 263: 35–70.

————, B. Grant, F.M.M. Clarke and T. Asami. 1994. A long-term assessment of *Biston betularia* (L.) in one U.K. locality (Caldy Common near West Kirby, Wirral), 1959–1993, and glimpses elsewhere. *Linnean* 10 (2): 18–26.

————, G.S. Mani and G. Wayne. 1985. Evolution in reverse: clean air and the peppered moth. *Biol. J. Linn. Soc.* 26: 189–99.

Cook, L.M. 2000. A century and a half of peppered moths. *Entomologist's Record* 112: 77–82.

————. 2000. Changing views on peppered moths. *Biol. J. Linn. Soc.* 69: 431–41.

————, R.L.H. Dennis and G.S. Mani. 1999. Melanic morph frequency in the peppered moth in the Manchester area. *Proc. Roy. Soc. Lond. B* 266: 293–7.

———— and G.S. Mani. 1980. A migration-selection model for the morph frequency variation in the peppered moth over England and Wales. *Biol. J. Linn. Soc.* 13: 179–98.

————, K.D. Rigby and M.R.D. Seward. 1990. Melanic moths and changes in epiphytic vegetation in north-west England and north Wales. *Biol. J. Linn. Soc* 39: 343–54.

Coyne, Jerry A. 1998. Not black and white. *Nature* 396: 35–6. 5 November 1998.

Creed, E.R. 1971. Melanism in the two-spot ladybird, *Adalia bipunctata*, in Great Britain. In E.R. Creed (ed.), *Ecological Genetics and Evolution*, pp. 152–74.

——, D.R. Lees and M.G. Bulmer. 1980. Pre-adult viability differences of melanic *Biston betularia* (L) (Lepidoptera). *Biol. J. Linn Soc.* 30: 31–44.

Demuth, R.F. 1979. Obituary: Dr H.B.D. Kettlewell. *Entomologist's Record*. 1/X/79.

Edleston, R.S. 1864. *Amphydasis betularia*. *Entomologist* 2: 150.

Fisher, R.A. 1927. On some objections to mimicry theory: statistical and genetic. *Transactions of the Entomological Society of London* 75: 269–78.

—— and E.B. Ford. 1947. The spread of a gene in natural conditions in a colony of the moth *Panaxia dominula*. *Heredity* 1: 143–74.

—— and E.B. Ford. 1950. The 'Sewall Wright' effect. *Heredity* 4: 117–19.

Ford, E.B. 1937. Problems of heredity in the Lepidoptera. *Biological Review* 12: 461–503.

—— and J.S. Huxley. 1927. Mendelian genes and rates of development in *Gammarus chevreuxi*. *British Journal of Experimental Biology* 5: 112–34.

—— and H.B.D. Kettlewell. The Experimental Study of Evolution/The Experimental Study of Evolution: A Further Discussion (audiotape).

Gould, Stephen Jay. 1983. The hardening of the Modern Synthesis. In Marjorie Grene (ed.), *Dimensions of Darwinism: Themes and Counterthemes in Twentieth-Century Evolutionary Theory*.

—— and R.C. Lewontin. 1979. The spandrels of San Marco and the Panglossian paradigm: a critique of the adaptationist program. *Proc. Roy. Soc. Lond.* B 205: 581–98.

Grant, B.S. 1999. Fine turning the peppered moth paradigm. *Evolution* 53: 980–4.

——, D.F. Owen and C.A. Clarke. 1996. Parallel rise and fall of melanic peppered moths in America and Britain. *Journal of Heredity* 87: 351–7.

——, A.D. Cook, C.A. Clarke and D.F. Owen. 1998. Geographic and temporal variation in the incidence of melanism in peppered moth populations in America and Britain. *Journal of Heredity* 89: 465–71.

—— and R.J. Howlett. 1988. Background selection by the peppered moth (*Biston betularia* Linn.): individual differences (Lepidoptera: Geometridae). *Biol. J. Linn. Soc.* 33: 217–32.

Hailman, J.P. 1982. Evolution and behaviour: an iconoclastic view. In H.C. Plotkin (ed.), *Learning, Development, and Culture*, pp. 205–54. London: John Wiley and Sons.

Haldane, J.B.S. 1924. A mathematical theory of natural and artificial selection. *Transactions of the Cambridge Philosophical Society* 23: 26.

——. 1956. The theory of selection for melanism in the Lepidoptera. *Proc. Roy. Soc. B* 144: 217–20.

——. 1957. The cost of natural selection. *J. Genet.* 55: 511–24.

Harrison, J.W.H. 1927. The induction of melanism in the Lepidoptera and its evolutionary significance. *Nature* 119: 127–9.

——. 1956. Melanism in the Lepidoptera. *Entomologist's Record and Journal of Variation* 68: 172–81.

Holdrege, Craig. Spring 1999. The case of the peppered moth illusion. *Whole Earth.*

Howlett, R.J. and M.E.N. Majerus. 1987. The understanding of industrial melanism in the peppered moth (*Biston betularia*) (Lepidoptera: Geometridae). *Biol. J. Linn. Soc.* 30: 31–44.

Jones, David A. 1989. 50 years of studying the scarlet tiger moth. *Trends in Ecology and Evolution* 4 (10): 298–301.

Jones, J.S. 1982. More to melanism than meets the eye. *Nature* 300: 109–10.

Kettlewell, H.B.D. 1955a. Recognition of appropriate backgrounds by the pale and black phases of Lepidoptera. *Nature* 175: 943–4.

———. 1955b. Selection experiments on industrial melanism in the Lepidoptera. *Heredity* 9: 323–42.

———. 1956a. Further selection experiments on industrial melanism in the Lepidoptera. *Heredity* 10: 287–301.

———. 1956b. Melanism and an answer to Heslop-Harrison. *Entomologist's Record and Journal of Variation* 68.

———. 1956c. An answer to one thought on reading Dr Ford's book *Moths. Entom. Rec. and J. Var.* 68: 286–92.

———. 1958. A survey of the frequencies of *Biston betularia* L. (Lep.) and its melanic forms in Britain. *Heredity* 12: 51–72.

———. 1959. Darwin's missing evidence. *Scientific American* 200: 48–53.

———. 1965. A 12-year survey of the frequencies of *Biston betularia* L. and its melanic forms in Great Britain. *Entom. Rec. and J. Var.* 77: 195–218.

——— and D.L.T. Conn. 1977. Further background-choice experiments on cryptic Lepidoptera. *J. Zool.* (London) 181: 371–6.

Kimler, W.C. 1983. Mimicry: views of naturalists and ecologists before the modern synthesis. In Marjorie Grene (ed.), *Dimensions of Darwinism*, pp. 97–127.

Kimura, M. 1968. Evolutionary rate at the molecular level. *Nature* 217: 624–6.

King, J.L. and T.H. Jukes. 1969. Non-Darwinian evolution: random fixation of selectively neutral mutations. *Science* 164: 788–98.

Kingsland, Sharon E. 1986. Mathematical figments, biological facts: population ecology in the Thirties. *Journal of the History of Biology* 19 (2): 235–56.

Lambert, D.M., C.D. Miller and T.J. Hughes. 1986. On the classic case of natural selection. *Rivista di Biologia* 79: 11–49.

Lees, D.R. 1971. The distribution of melanism in the pale brindled beauty moth, *Phigalia pedaria*, in Great Britain. In E.R. Creed (ed.), *Ecological Genetics and Evolution*, pp. 152–74.

———. 1981. Industrial melanism: genetic adaptation of animals to air pollution. In J.A. Bishop and L.M. Cook (eds), *Genetic Consequences of Man-made Change*, pp. 129–76.

———. 1979. Obituary: H.B.D. Kettlewell. *Entomologist's Monthly Magazine*: 251–5.

——— and E.R. Creed. 1975. Industrial melanism in *Biston betularia*: the role of selective predation. *Journal of Animal Ecology* 44: 67–83.

Leigh, Egbert Giles. 1986. Ronald Fisher and the development of evolutionary theory. I. The role of selection. *Oxford Surveys in Evolutionary Biology*, vol. 3. R. Dawkins and M. Ridley (eds). Oxford: Oxford University Press.

Lewontin, R.C. 1972. Testing the theory of natural selection. *Nature* 236: 181–2.

———. 1987. Polymorphisms and heterosis: old wine in new bottles and vice versa. *Journal of the History of Biology* 20(3): 337–49.

——— and J.L. Hubby. 1966. A molecular approach to the study of genic heterozygosity in natural populations. II. Amount of variation and degree of heterozygosity in natural populations of *Drosophila pseudoobscura*. *Genetics* 54: 595–609.

Liebert, T.G. and P.M. Brakefield. 1987. Behavioral studies of the peppered moth *Biston betularia* and a discussion of the role of pollution and lichens in industrial melanism. *Biol. J. Linn. Soc.* 31: 129–50.

Mani, G.S. 1982. A theoretical analysis of the morph frequency variation in the peppered moth over England and Wales. *Biol. J. Linn. Soc.* 17: 259–67.

———. 1990. Theoretical models of melanism in *Biston betularia* – a review. *Biol. J. Linn. Soc.* 39: 355–71.

Manley, T.R. 1981. Frequencies of the melanic morph of *Biston cognataria (Geometridae)* in a low-pollution area of Pennsylvania from 1971 to 1978. *Journal of the Lepidopterists' Society* 35: 257–65.

Matthews, R. 1999. Scientists pick holes in Darwin moth theory. *The Daily Telegraph.* 18 March 1999.

Mayr, Ernst. 1992. Controversies in retrospect. *Oxford Surveys in Evolutionary Biology* 8: 1–30. Oxford: Oxford University Press.

Mikkola, K. 1979. Resting site selection by *Oligia* and *Biston* moths (Lepidoptera: Noctuidae and Geometridae). *Ann. Entomol. Fenn.* 45: 81–7.

———. 1984. On the selective force acting in the industrial melanism of *Biston* and *Oligia* moths (Lepidoptera: Geometridae and Noctuidae). *Biol. J. Linn. Soc.* 21: 409–21.

Owen, D.F. 1962. The evolution of melanism in six species of North American moths. *American Naturalist* 95: 227–33.

Provine, William P. 1990. Discussion: Population Genetics. *Bulletin of Mathematical Biology* 52 (1/2): 201–7.

———. 1992. The R.A. Fisher–Sewall Wright controversy. In Sahotra Sarkar (ed.), *The Founders of Evolutionary Genetics*, pp. 201–29. Netherlands: Kluwer Academic Publishers.

Rudge, David Wyss. 1999. Taking the peppered moth with a grain of salt. *Biology and Philosophy* 14: 9–37.

———. 2000. Does being wrong make Kettlewell wrong for science teaching? *Journal of Biological Education* 35(1): 5–11.

Sargent, T.D. 1966. Background selections of geometrid and noctuid moths. *Science* 154: 1674–5.

———. 1968. Cryptic moths: effects on background selections

on painting the circumocular scales. *Science* 159: 100–01.

———. 1969. Background selections of the pale and melanic forms of the cryptic moth, *Phigalia titea* (Cramer). *Nature* 222: 585–6.

———. 31 May 1974. Review: The Evolution of Melanism by Bernard Kettlewell. *Journal of the Lepidopterists' Society* 28 (2): 176–8.

———. 1983. Melanism in *Phigalia titea* (Cramer) (Lepidoptera: Geometridae): a fourteen-year record from central Massachusetts. *Journal of the New York Entomological Society* 91 (1): 75–82.

———. 1985. Melanism in *Phigalia titea* (Cramer) (Lepidoptera: Geometridae) in southern New England: a response to forest disturbance? *Jour. NY Ent. Soc.* 93(3): 1113–20.

———. 1990. Industrial melanism in moths: a reassessment. In M. Wiksten (compiler), *Adaptive Coloration in Invertebrates: Proceedings of a Symposium Sponsored by the American Society of Zoologists*, pp. 17–30. Texas A & M University, College Station, Texas.

———, Craig D. Millar and David M. Lambert. 1998. The 'classical' explanation of industrial melanism: assessing the evidence. In Max K. Hecht *et al.* (eds), *Evolutionary Biology*, vol. 30. New York: Plenum Press.

Shapiro, Arthur M. 2000. Review of *Melanism: Evolution in Action* by Michael E.N. Majerus. *Journal of the Lepidopterists' Society* 45(1): 38.

Sheppard, P.M. 1973. Pigmentation and Evolution. *Nature* 246: 535–536. December 21/28 1973.

Smocovitis, Vassiliki Betty. 1994. Organizing evolution: founding the Society for the Study of Evolution (1939–1950). *Jour. Hist. Biol.* 27 (2): 241–309.

———. 1999. The 1959 Darwin Centennial Celebration in America. *Osiris* 14: 274–323.

Sulloway, Frank J. 1979. Geographic isolation in Darwin's thinking: the vicissitudes of a crucial idea. *Studies in the History of Biology* 3: 23–65.

Turner, John R.G. 1977. Sheppard, Philip MacDonald. In Frederic L. Holmes (ed.), *Dictionary of Scientific Biography*, vol. 18 (II), pp. 814–16. New York: Charles Scribner's Sons.

———. 1980. Kettlewell, Henry Bernard Davis. In F. Holmes (ed.), *Dictionary of Scientific Biography*, pp. 459–71. New York: Charles Scribner's Sons.

———. 1985. Fisher's evolutionary faith and the challenge of mimicry. In *Oxford Surveys in Evolutionary Biology*, vol. 2, R. Dawkins and M. Ridley (eds), pp. 159–96. Oxford: Oxford University Press.

———. 1992. Stochastic processes in populations. In R.J. Berry, T.J. Crawford and G.M. Hewitt (eds), *Genes in Ecology*, op. cit.

Waddington, C.H. 1956. Genetic assimilation of the Bithorax phenotype. *Evolution* 10: 1–13.

West, D.A. 1977. Melanism in *Biston* (Lepidoptera: Geometridae) in the rural central Appalachians. *Heredity* 39: 75–81.

Wilson, Bryan. Edmund Brisco Ford. Memorial address delivered in the chapel of All Souls College. 12 March 1988.

Yoon, Carol Kaesuk. 1996. Parallel Plots in Classic of Evolution. *The New York Times*, p. C-1. 12 November 1996.

Glossary

ADAPTATION: a characteristic of an organism that helps it to cope better with the conditions of its environment.

ALLELE (ALLELOMORPH): one of several forms of the same gene, differing by mutation of the DNA sequence.

ALLOZYME: one of several forms of an enzyme coded for by different alleles at the same genetic locus.

ARTIFICIAL SELECTION: human attempts to enhance natural traits by breeding selectively from those organisms that most strongly manifest those traits.

BALANCE HYPOTHESIS: the claim that within a population natural selection holds many different alleles in equilibrium.

BATESIAN MIMICRY: the resemblance of a harmless species (the 'mimic') to another species (the 'model') that is endowed with poison, bad taste, or a sting or bite that makes it unpleasant to predators.

BIOMETRICIANS: Darwinist scientists at the beginning of the twentieth century who, in opposition to the Mendelians, believed that inherited traits were blended.

CHROMOSOME: structure in the nucleus of a eukaryotic cell consisting of DNA molecules that contain the genes.

CLASSICAL HYPOTHESIS: the claim that little genetic variation exists within a population.

CLINE: a gradual geographical change in the frequencies of different genotypes (or phenotypes) of a species within an interbreeding population.

CRYPSIS: the resemblance, by virtue of colour or pattern, of an organism to its habitual background: camouflage.

DARWINISM: the belief that natural selection is the overwhelming causal factor in evolution.

DIRECTIONAL SELECTION: selection in which individuals at one end of the distribution of a trait (such as tallness) are at an advantage over other individuals.

DNA (DEOXYRIBONUCLEIC ACID): a type of nucleic acid, found in the nucleus of plant and animal cells, that contains the genetic information.

DOMINANT: the stronger of a pair of alleles, expressed as fully when present in a single dose (heterozygous) as in a double dose (homozygous). Opposite of *recessive*.

DROSOPHILA: the fruit fly, a favourite organism for the study of genetics.

ECLOSION: the metamorphosis of a pupa into an adult insect.

EUGENICS: the belief that the best way to improve mankind is through selective breeding.

EVOLUTION: the change in groups of organisms over time, so that descendants differ from their ancestors; the change in organisms brought about by adaptation, variation and differential survival/reproduction, by the process of natural selection.

FITNESS: the average contribution of one allele or genotype to the next generation or to succeeding generations, compared with that of other alleles.

FOUNDER EFFECT: a situation in which a very small number of individuals from a population establish a new colony whose genetic variability may be very different from that of the parental population.

GEL ELECTROPHORESIS: a technique for detecting variations in molecular genes by tracking their movement through a gel subjected to an electrical field.

GENE: the unit of heredity, today known to consist of a sequence of base pairs in a DNA molecule.

GENE POOL: the collective genes of a population or species.

GENETIC DRIFT: the claim that in small populations accidents of mating can outweigh the effects of selection.

GENOTYPE: the genetic constitution of an individual organism; often, its genetic composition at a specific locus singled out for discussion.

GERM CELLS: cells destined to become reproductive cells.

GRADUALISM: the belief that evolutionary change occurs gradually.

HABITAT: the specific type of local environment in which an organism lives.

HETEROZYGOTE: an individual organism that bears different alleles of a given gene at a particular locus.

HETEROZYGOUS ADVANTAGE (HETEROSIS): the situation in which the heterozygote is fitter than either homozygote.

HOMOZYGOTE: an individual organism that bears two copies of the same alleles at a given gene locus.

IMAGO: the adult stage of an insect.

INDUSTRIAL MELANISM: the observed increase in the incidence of melanism (melanic morphs) in many species, particularly bark-like cryptic moths, in heavily industrialized parts of the world.

INSTAR: the stage in an insect's life between two moults. An insect that has recently hatched from an egg and has not yet moulted is in its first instar. The final instar is the adult.

LAMARCKISM: a theory predating Darwin to explain evolution. Generally referred to as the inheritance of acquired characteristics through use and disuse of parts.

LARVA: the second stage in the life cycle of insects that undergo full metamorphosis; in Lepidoptera, caterpillars.

LOCUS: a site on a chromosome occupied by a specific gene; more loosely, the gene itself.

MACROEVOLUTION: evolution occurring above the species level, over long periods of time.

MACROMUTATION: mutation that causes major changes, a saltation.

MELANISM: the deposition of melanin pigment in a structure, rendering that structure blackish. In moths, generally refers to a blackening of the wings, usually controlled by a single gene.

MENDELISM: the genetical theory, based on Mendel's laws, according to which the units of inheritance do not change or blend when transmitted to offspring. Offspring can carry genes for traits that may not be expressed until a subsequent generation.

MICROEVOLUTION: evolution occurring at or below the species level, over short time periods.

MODIFIER GENES: a group of genes, each with small effects, that influences the expression of a major gene.

MOLECULAR BIOLOGY: the branch of biology that deals with organic processes at the level of molecules, as opposed to cells, organs, organisms, populations, species, and so on.

MORPH: form; a distinctive variant within a species that is characterized by genetic continuity from generation to generation, and is maintained in the species by some selective advantage.

MUTATION: a sudden change in the genetic material that determines a trait or set of traits of an organism. The mutation may be due to a change in an individual gene, in the number of chromosomes, or the structure of a chromosome.

NATURAL SELECTION: the principal mechanism of evolutionary change according to Charles Darwin. The mechanism whereby advantageous heritable traits, which enhance the ability of an organism to survive or reproduce, are more likely than disadvantageous traits to be passed on to future generations.

NEUTRAL MUTATION: a mutation whose expression has no fitness consequences; i.e., is selectively neutral.

NEO-DARWINIAN THEORY (SYNTHESIS): the later development of Darwin's evolutionary theory, from the 1930s to 1950s, incorporating Mendelian genetics and other newer findings.

ORTHOGENESIS: the belief that life develops according to a predetermined momentum, not subject to the influence of external factors. Also called *airistogenesis*.

PHENOTYPE: the physical and behavioural properties of an organism, manifested throughout its life.

PHEROMONE: a chemical substance secreted or released by an organism that influences the behaviour of other individuals of the species. For example, some pheromones function as sexual attractants.

POLYMORPHISM (GENETIC): the occurrence in a single inter-breeding population of two or more sharply distinct heritable forms, the rarest of which is too frequent to be maintained simply by recurrent mutations.

POPULATION: a group of organisms of the same species living and breeding together in the same place.

POPULATION GENETICS: the extension of Mendelian genetics to groups of organisms, showing the effects of selection, mutation and other factors on the gene pool of populations.

PUNCTUATED EQUILIBRIUM: the claim that evolution consists of long periods of stability (stasis) interrupted now and then by rapid major change.

PUPA: the life-cycle stage between larva and adult (imago) in insects having complete metamorphosis.

RECESSIVE: a recessive allele is expressed in the phenotype of homozygotes and not in that of heterozygotes. Opposite of *dominant*.

SALTATIONISM: the claim that evolutionary change occurs through sudden jumps, or macromutations.

SELECTIONISTS: evolutionists who hold that most evolutionary change results from the action of natural selection on small heritable differences.

SPECIATION: the process whereby new species form.

SPECIES: groups of interbreeding or potentially interbreeding populations that are reproductively isolated from other such groups.

STABILIZING SELECTION: selection in which individuals intermediate in the distribution of a trait are at an advantage over those at either extreme.

SUGARING: a technique for attracting certain moths, usually by applying a sugary and slightly alcoholic mixture to tree trunks.

SYSTEMATICS: the scientific theory behind the classification of organisms.

TAXONOMY: the study of the classification of organisms.

TELEOLOGY: the study of final causes; the quality of being directed toward a definite end or having a definite purpose.

ULTRA-ADAPTATIONISM: the belief that everything has an adaptive function.

VITALISM: the claim that organisms are driven by non-material life forces.

Chronology

1848 R.L. Edleston observes a black mutant form of the peppered moth *Biston betularia* in Manchester, England.

1859 *On the Origin of Species* is published.

1896 J.W. Tutt proposes that industrial melanism is a vivid example of Darwinian natural selection.

1900 Industrial melanism is singled out by the Evolution Committee of the Royal Society of London as a matter of urgent scientific importance.

1924 J.B.S. Haldane calculates the selective advantage of the melanic peppered moth at 50 per cent.

1927 John W. Heslop Harrison proposes an environmental induction model of industrial melanism.

1937 In the first of a series of papers, Edmund Brisco Ford revives Tutt's Darwinian theory of industrial melanism.

1953 H.B.D. Kettlewell puts the industrial melanism hypothesis to an experimental test near Birmingham, England.

1955 Kettlewell repeats his Birmingham experiment and performs another experiment in an unpolluted wood in Dorset. Niko Tinbergen films birds eating moths.

1959 'Darwin's Missing Evidence' by H.B.D. Kettlewell is published in *Scientific American*, launching the famous catchphrase.

1973   *The Evolution of Melanism* by Bernard Kettlewell is published.

1979   H.B.D. Kettlewell dies.

1998   *Melanism: Evolution in Action* by Michael E.N. Majerus published. A book review in *Nature* asserts that 'we must discard *Biston* as a well-understood example of natural selection in action . . .'

Picture Credits

Index